# *The Tyranny of Survival*

# THE TYRANNY OF SURVIVAL

## And Other Pathologies of Civilized Life

Daniel Callahan

UNIVERSITY PRESS OF AMERICA

LANHAM • NEW YORK • LONDON

**University Press of America,® Inc.**

4720 Boston Way
Lanham, MD 20706

3 Henrietta Street
London WC2E 8LU England



Printed in the United States of America



Originally published by Macmillan Publishing Co., Inc.

Co-published by arrangement with the
Center for the Study of Values,
University of Delaware

**Library of Congress Cataloging in Publication Data**

Callahan, Daniel, 1930-
The tyranny of survival, and other pathologies of civilized life.

"Co-published by arrangement with the Center for the Study of Values, University of Delaware"—T.p. verso.
Originally published: New York : Macmillan, 1973.
Includes index.
1. Civilization, Modern—1950- . 2. Technology and civilization. 3. Technology—Moral and ethical aspects. I. Title.
CB428.C3 1985 909.82 85-5316
ISBN 0-8191-4636-6 (pbk. : alk. paper)

# Acknowledgments

PARTS OF THIS BOOK have appeared in other forms in other places. The classifications of technology which appear in Chapter 3 were first developed as a background paper prepared for the U.S.A. Task Force on the Future of Mankind and the Role of the Churches in a World of Science-Based Technology. An earlier version of Chapter 4 appears in Philip Wogaman (ed.), *The Population Crisis: Moral Analysis of Policy Proposals* (Washington, D.C.: Public Affairs Press, 1973). The first part of Chapter 6 first appeared in *The Humanist*, Vol. 32 (Sept./Oct. 1972), under the title of "Normative Ethics and Public Morality in the Life Sciences"; and the second part was prepared for the 1972 Annual Meeting of the Institute of Society, Ethics and the Life Sciences (of which a condensed version was published in *The Center Magazine*, Vol. 5 [July/Aug. 1972] under the title of "Living with the New Biology"). A major portion of Chapter 7 represents a reworking of a paper originally prepared for The Population Council ("Ethics and Population Limitation," An Occasional Paper [New York: The Population Council, 1971], a somewhat different version of which was also published in *Science* Vol. 175, pp. 487–494 (Feb. 4, 1972), copyright 1972 by the American Association for the Advancement of Science. Part of Chapter 8 originally appeared in Bruce Hilton, Daniel Callahan *et al.* (editors), *Ethical Issues in Human Genetics: Genetic Counseling and the Use of Genetic Knowledge* (New York: Plenum Press, 1973).

In addition, excerpts from *Freud: The Mind of the Moralist* by Philip Rieff, copyright © 1959 by Philip Rieff, are reprinted by permission of The Viking Press, Inc. Excerpts from *Civilization and Its Discontent* by Sigmund Freud, translated and edited by James Strachey, copyright © 1961 by James Strachey, are reprinted by permission of W. W. Norton & Company, Inc. Excerpts from *The Triumph of the Therapeutic* by Philip Rieff is reprinted by permission of Harper & Row, Publishers, Inc.

FOR

*Willard Gaylin*

AND

*Leon R. Kass*

WITHOUT WHOM I COULD NOT

HAVE DONE THIS BOOK AT ALL,

BUT COULD ALSO HAVE DONE IT

MUCH FASTER

# *Contents*

# *Preface*

A BRIEF EXPLANATION is in order about the background and plan of this book. For some years, I have been intrigued with Freud's *Civilization and Its Discontents*, particularly by his posing of the conflict between the individual and civilization as an inherent part of the human condition. At one time I projected a study of Philip Rieff, Norman O. Brown and Herbert Marcuse, each of whom has been no less captivated by Freud's harsh rendering of the reality principle, and each of whom (save Rieff) proposed a way out of a situation which Freud claimed admits of no escape. Of the three, however, Rieff's writing made the most lasting impression. Norman O. Brown, as *Love's Body* finally made clear, has nothing to offer but a regression to nothingness; nice, but Herbert Marcuse, particularly the Marcuse of *Eros and Civilization*, took the Freudian argument in a promising direction; that is a book of enduring value, not to be confused with Marcuse's later, less successful, venture into radical politics. He should, like Freud, have less hope for political solutions to enduring human dilemmas.

Rieff has held my attention the longest because he is the

least hopeful of the three, the hardest on contemporary culture, the most bitingly and sometimes bitterly ironic in his analyses of the personal "liberation" movements which abound in our society. As a good, though ambivalent, disciple of Freud he is properly skeptical of any and all promises of a final freedom. He is also a most astute critic of contemporary individualism, far harsher and more penetrating than those who emphasize the political and economic liabilities of individualism; he refuses to provide us with the social plans and revolutions which ideologues have so readily available for the salvation of us all.

Not everyone finds these characteristics of Rieff endearing. I have spent a number of years trying to persuade people to read him, particularly *The Triumph of the Therapeutic.* In the late '60s, I had little success. Many who tried found him unintelligible or unpalatable, or both. Those were not the years to be putting the heat on self-realization and personal fulfillment. Many of my friends thought he simply failed to understand how many fragrant new winds of cultural rejuvenation were blowing. He was out of touch, they said, with the new movements toward community, with a younger generation throwing off the shackles of narrow liberal self-interest, with everything Charles Reich was later to christen "Consciousness III." He was deflating those pretensions before they had a chance to get off the ground, and that was unforgivable in those days.

Times have changed, not so much on the surface where the business and the self-realization industries are making fat profits on Con III merchandise, but down there very deeply where many of us find it hard to keep whistling in the dark. *The Triumph of the Therapeutic* reads as well in 1973 as it did in 1966, and I find now that people say they like the book, something few (other than reviewers) said when it was first published.

## *Preface*

Freud and Rieff have been on my mind almost constantly in the past few years. In part, that was because they have both spoken to one of the most fundamental human problems: how are people to live, and how are they to live together. One could hardly exist in our culture and not be forced to give some thought to those questions. Freud and Rieff make fine guides, as long as one is congenial with ascetics and anti-prophets. In part also, these same years saw the development of the Institute of Society, Ethics and the Life Sciences. From the first my colleagues and I have tried to work through some of the nasty ethical problems which are raised by advances in medicine and biology, those two preeminent scientific-technological disciplines of the '70s. All the while, however, I have been left with a feeling of discomfort. It has proved exceedingly difficult to deal adequately with the larger cultural context within which the ethical dilemmas appear; it is far easier to bring to the ethical problems the traditional analytic tools of the philosopher, the lawyer and the social scientist—they can be made more or less tractable that way. But that does not deal effectively with what Rieff has called "the unwitting part" of a culture, that part which shapes the attitudes, images and expectations that more often than not form the psychological and cultural roots of particular moral principles and ethical proclivities.

This book is an attempt to make use of Freud's and Rieff's insights into the dynamic of personality and culture in order to understand our problems with technology, particularly the technologies of medicine and biology. Freud has convinced me that the conflict between the individual and civilization admits of no solution, a critical insight for any attempt to balance the interest of individuals and of society in the use of technology. Rieff has persuaded me that the most dangerous cultural disease of our society is individualism, which leads people to think they can and must have everything they desire. The mixture of individualism and technological *hubris* is about as dangerous a mix as can be imagined. And I have convinced myself that the heated cry for survival, especially survival of the species, is probably as dangerous as that of individualism (its mirror-image). Put individualism, technology and an obsession with survival together—that is when the whole house of cards will burn down. Yet with Freud I want to defend technology and the need for some degree of individual freedom and communal endurance and unity. A great part of this book represents an effort to see if some kind of sensible settlement can be reached among these opposed insights and impulses. I use the problem of population growth and that of the prospect, rapidly being realized, of genetic counseling and engineering, as case studies.

Chapter 1 establishes the framework for my investigation, laying a particular stress on the need for viable reality principles in dealing with technology. Chapter 2 is an analysis of *Civilization and Its Discontents,* particularly the structure of Freud's argument about the relationship between the individual and civilized society. Chapter 3 sets forth a classification scheme for understanding the origins of technology and the different human needs which technology serves (and often dis-serves). Chapter 4 analyzes the notion of "survival," contending that while it is obviously of fundamental importance (try living without it) it can be and has been transmuted into the grossest forms of tyranny. Chapter 5, which attempts to understand the writings of Philip Rieff on individualism, on "psychological man," also points out that individualism in technological societies can become as great a source of psychological tyranny as a single-minded dedication to group survival; in fact, that chapter argues, there is an intimate connection between the dual concern for survival and individualism which is one of the striking marks of our time. Chapter 6 moves the argument to another plane, contending that both "normative ethics" and a "public morality" are required if there is to be any possibility of moving beyond the ethical geniality which is the mark of an individualistic, release-bent society. Chapter 7 focuses the discussion on population growth and its relationship to technology and survival. Chapter 8 takes up genetic counseling and engineering, noting that neither individualism nor survivalism provide adequate norms with which to assess advances in genetic knowledge and technology. Chapter 9 develops some of the important ingredients which must go into a science of technological limits; it says Yes to a controlled technology (indispensable for human welfare) and No to a technology which knows no prohibitions and refusals.

Though I have talked about parts of this book on many occasions with my colleagues at the Institute, there are other parts which will come as something of a surprise to them. I delight in the help and counsel they have given me on those parts with which they are familiar (especially Willard Gaylin, Robert M. Veatch, Marc Lappé, Peter Steinfels, Robert C. Neville and Leon R. Kass); and I do not hold them accountable for my errors and excess. But I also delight in the fact that there are parts of this book I have talked with no one about; one hopes, now and then, to have a few surprises left. I hope they—and my readers—find them tolerable.

I would particularly like to thank Drew Christianson and my wife Sidney for their careful reading of the manuscript, and Lyn Brydson and Ellen McAvoy for their help in seeing the manuscript through the tedium of typing.

—DANIEL CALLAHAN

*Institute of Society, Ethics and the Life Sciences*
*Hastings-on-Hudson, New York*

# CHAPTER 1

# *Technology and the Reality Principle*

IN HIS SHORT MASTERPIECE *Civilization and Its Discontents,* Sigmund Freud predicted that someone would eventually undertake research into the "pathology of civilized communities."[1] The primary symptoms of the pathology are those "social neuroses" which make life in civilized societies a burden and a threat, particularly war and aggression. Freud did not underestimate the complexity of the research he called for. "The diagnosis of collective neuroses, moreover," he observed, "will be confronted by a special difficulty. In the neurosis of an individual we can use as a starting point the contrast presented to us between the patient and his environment which we assume to be 'normal.' No such background as this would be available for any society similarly affected; it would have to be supplied in some other way."[2] In short, when we move from the level of the individual to that of society, there exists no readily available standard of "normalcy," or, as I would prefer to put it, no available normative standard of a proper and sane civilized society. To be sure, it is possible to sketch the rudiments of such a standard, and terms such as "peace," "justice," "freedom," "domestic tranquility," and the like might be part of any formal definition.

Freud, however, posed a harder problem: how are we to understand the persistent gap, not only in our Western culture but in *all* historical cultures, between our dreams of civilized "happiness" and the doleful reality of life together? If what I have called a normative standard for civilized life is to have any useful purpose, it must do more than simply propose a transcendence of the gap. That has never been possible, however much utopians and millenarians have helped to keep alive or extend the stuff of visions. Before a standard can be developed, the "pathology of civilized communities" must go forward.

The purposes of this study, while not altogether modest, are still more modest than the kind of research Freud envisioned. A full pathology is beyond me. I do not know why human beings go to war or why, some of the time, I hate my neighbor. What will concern me is a problem no less complicated, however. Why does technology, thought to provide happiness, manage nonetheless to make its own distinctive contribution to the "pathology"? That subject is not exactly a fresh one. We have heard more than a little in recent years about the harmful by-products of technology, especially those which have allowed human beings to split the atom and exploit the natural resources of the earth. In those instances, it has become easy and legitimate game to point to the real or possible disasters. Yet it has proved far more difficult to cope with those technologies where the evidence of harm is still mixed. This is particularly the case with medical and biological technologies. By and large, the general wariness about technology has not extended into those domains where the good of the human body is concerned. The "conquest" of disease goes forward, even if now and then a doubting voice is raised about the desirability of such exotica as cloning and test-tube babies.

I will dwell on two technological case histories in this book, that of population growth, on the one hand, and genetic counseling and engineering, on the other. Between them they circumscribe what is by now a central question, the extent to which mankind can or must control the biological quantity and the genetic quality of life. The question could not arise but for technological developments. World population growth is the result of those medical advances which have radically reduced deathrates (without a corresponding drop in birthrates). The issue of genetic quality has arisen because of the increased understanding of genetic mechanisms coupled with the possibility of more sophisticated means of medically intervening into and manipulating those mechanisms, whether directly or indirectly.

There are two senses in which these problems raise the issue of a possible pathological condition. The first is that of unexpected and undesired by-products of technological development. In the case of population growth, that by-product is the world population growth of 2 percent per annum, which at present rates will lead to a doubling of world population in some thirty-seven years. The cause of this growth—the reduction in deathrates—would in ordinary cases be considered an unmitigated good. Who wants to die, or see their children die before them? But this is no "ordinary case," it turns out. The irony is that the very success of medical technology in controlling death at present may have set the stage for even more deaths in the future. In the case of genetic control, we have yet to see what the by-product of genetic manipulation will be. We only know that the primary motivation for this manipulation stems from a desire to reduce the frequency of genetic disease and the suffering it carries with it. Yet there is good reason to believe that this manipulation could have some undesirable side-effects as well, social and biological.

Up to this point, my discussion of technology is on by-now

familiar grounds: technology as a double-edged sword, capable of producing good at the price of creating new evils. If there is a pathological condition here, it may stem either from an unwillingness of human beings to take account of the long-term evils which their strategies for fulfilling a short-term desire for happiness can bring, and thus to take steps to calculate balances; or from some innate impossibility of realizing all human good at the same time—a fateful inevitability in the nature of things human, which counters every move toward good with a countermove toward evil. There is, however, a second sense in which the problems of population growth and genetic quality bespeak a possible pathological condition. I am thinking of those "civilized" societal steps which modern human beings take, or have considered taking, in order to cope with or exploit either the fruits of their scientific knowledge or the unwanted by-products and unintended consequences of their technological developments.

A distinction must be made between controlling population growth and control of genetic quality; some important differences quickly appear. In the instance of population growth, the problem itself directly stems from technology and already exists. The response has been a twofold strategy, involving efforts to develop safer, cheaper and more effective contraceptives and abortifacients (a technological response to a technologically induced problem), and to find ways of changing social attitudes and institutions to promote controlled fertility and smaller families. At no point has there been a move to reverse the downward trend of deathrates; efforts have concentrated solely on reduction of birthrates. The potentially pathological nature of these latter efforts stems from the fact that, in order to effectively reduce birthrates, it may be necessary for society by manipulation or violence to intervene directly into individual procreative behavior. If people will not voluntarily have smaller families, it may be necessary to force them to do so; and, for that matter, it is possible by subtle means to manipulate them to the point that they "voluntarily" choose smaller families. One way or the other, the stage is set for society to take aggressive action against those who, producing a large family, pose a threat to the social good.

A different situation appears with control of genetic quality. The technology is in its infancy, basic knowledge of genetics still sketchy and, with the exception of prenatal diagnosis and amniocentesis (useful in diagnosing only a limited range of defects), there exist few medical tools to allow individuals or society to have any control over genetic quality. Thus it is not yet possible to speak of any known evils which will *inevitably* result from greater genetic knowledge and more sophisticated clinical applications. Moreover, there exists enough knowledge and technological "promise" to stimulate enormous scientific efforts toward control of genetic quality. That the emphasis has so far fallen most heavily on ridding mankind of genetic disease should not obscure the fact that the vision of genetic "improvement" has a lively life just below the surface, in the stirrings of a new eugenics movement. Assume that it will become possible to control genetic quality, particularly through control or elimination of deleterious genes. If it is known (as it is) that the social cost of genetic disease is high, then it may seem only logical at that point—even humane—to contend that individuals, once they have the necessary knowledge and means, have an obligation not to produce defective children, and that society has the right to make certain that they do not. Would the resultant coercion be an instance of a social pathology, one caused by pressure to use a technology because it was there to be used?

But what are we to count as a pathological condition? I mean this question in both of the senses of "pathology" used above. First, what kinds of technological developments introduce pathological by-products and why, and to what extent and in what ways do we want to accept, live with and manage those by-products? In terms of population growth, how willing should we be to see a further control of death and, with it, a continuing population problem? In terms of genetic quality, how willing should we be to intervene consciously into the human genetic constitution and thus possibly run the risk of unsettling the results of a long human evolutionary history which has demonstrated a powerful survival strength? If there are some known risks, or if risks should appear, by what ethical norms are we going to judge promise and risk? Second, in either trying to correct some harmful consequences of technol ogy, or in trying to decide how far to make a greater use of technology, to what extent will social pathologies—or, to use Freud's term, "social neuroses"—be introduced into society? And if they will be, how are we to judge whether they will be worth the technological gain?

The point of using population and genetics as case histories of technologically generated problems is that both bring out, in a highly acute, painful way, some of the most inherently difficult dilemmas of technological change. The most serious difficulty in weighing the advantages and disadvantages of some technologies is that, however much they portend or display real harms, the good they promise seems undeniably desirable. It is impossible to escape the dilemma unless the good in view is forgone altogether. In principle it would be possible at least to ameliorate the population problem by allowing deathrates, particularly infant mortality rates, to rise; by refusing, in a word, to make use of (or extend any further) available life-saving medical technologies. Genetic quality would pose no dilemma if individual couples could give up their desire to have healthy children, or if no further efforts were possible to apply genetic knowledge to the cure of existing diseases. But human beings have shown little tendency to acquiesce in the reality of death, particularly the death of their children, nor have they found it a joy to suffer for and with a genetically defective child.

One must of course eventually face the question of whether, in the end, death itself should be indefinitely combated. Is death a friend or an enemy? No less must one face the question of whether everything we are prone to call evil—genetic disease, for example—is in fact an unmitigated evil. But I am going to take it for granted, at the outset, that it is morally legitimate to seek the good of life rather than the darkness of death, and the blessing of a sound body rather than one which is deformed. The promise of science and technology is that it can assist the attempt to gain those goods and blessings. No other way has been found.

But it is patently clear at this point in history that a price will be paid for the good which technology delivers. I am not saying that a price *must* be paid. All we know is that time and again with frustrating regularity there seems to be a price. No one could have guessed that with the introduction of modern sanitation, and after that modern antibiotics and pesticides, the result would have been a population crisis. But even if that had been known, it is difficult to imagine that efforts to control killing diseases would have been curtailed. We must seek life, and that means we must seek to stay the hand of death—even if, eventually, death will return again. That the ills brought on by some technologies are indirect (death control leading to population growth), while those brought on by others may be direct (control of individual genetic quality leading to a decline in species genetic quality), introduces an important distinction for the purpose of analyzing technology. But at the receiving end, in the life of human beings, this may be a distinction without a difference if one is dead or sick either way.

The inevitable issue which finally emerges is that of what

is "good" for human beings, as individuals and as societies. That is a question of value, and as Freud observed, ". . the judgments of value made by mankind are immediately determined by their desires for happiness."[3] Freud, though, went on in the same passage to stipulate what he meant: "in other words, that those judgments are attempts to prop up their illusions with arguments." I do not share his sweeping judgment. That human beings desire happiness can hardly be doubted, just as it is evident they are frequently wrong about what will bring them happiness. But it has not been shown, by Freud or anyone else, that all judgments on what will bring happiness are no more than rationalizations. Freud himself did not accept war as a human good; and it is reasonable to assume he felt that the absence of war would do no great harm to the desire for happiness. But of course these are easy retorts and Freud was pressing a more fundamental point. He was judging the whole quest for happiness to be an illusion. He had, he said, no consolation to offer those who still think (as ever they will) that happiness might be found.

The case of technology offers some strong evidence in support of his pessimism. The investment of human hope in technology has been enormous, from the satisfaction of basic physical needs to the fulfillment of the most vagrant whims and desires. The success stories of technology have been no less enormous, giving considerable historical substance to the hopes. Nonetheless, even if one puts aside the patently harmful effects of technology, there is little evidence that people have become happier or more satisfied. They are healthier, live longer and more comfortable lives, can travel farther and faster, know more—and yet evidence of dissatisfaction is rampant. However much life "improves" it has not been enough to make happiness still seem anything more than a vague and distant dream, an illusion. It is easy enough, to be sure, to play the romantic, pointing to the apparent folly lurking behind the belief that technology can deliver "progress" and happiness. One implication of that stance is always that life was better or at least no worse in simpler times, when people had a more acute sense of the limitations of human possibility. But the eagerness with which all cultures have embraced and continue to embrace technology, once they have been given even the slightest chance of doing so, should indicate that pre- or non-technological people and societies do not find their lives altogether satisfying. They too want something more, and so far as they can see technology offers them as great a chance for this "more" as anything else they have been offered.

That human beings can neither achieve happiness by means of technology nor give up their hope in technology points to a profound riddle of human nature. To use Freud's term, is there a "reality principle" which should serve as a check to investment in technology? If one could be discovered, then boundaries might be set to hope. Technology would be asked to do only that which is possible, and other means could be sought to fulfill hopes which outrun technological possibilities.

One element of a reality principle has, by now, been recognized: technology can produce harmful by-products. The case of DDT is one of the many classic stories, wiping out malaria but in the process doing irreparable harm to the environment. The basic premise of the ecological movement has been the necessity of protecting nature—and thus the source of man's physical nurture—from the harm which can be done by an unregulated, technological exploitation. Yet this insight has only a limited application. It does not show that technology *cannot* bring happiness, but only that it will not do so if it is used in ignorance of the balance of nature. It is an insight perfectly compatible with a conviction that man's best hope still resides in technology—only a bit of humility, and a better understanding of scientific and ecological principles is necessary. This has been a common response to technological disasters. That some elements of the ecology movement have called into question the attempted dissociation of man and nature, often an implicit tenet of the faith in technology, presses the whole issue into an important new area, toward what could eventuate into a new reality principle. Still, this represents a fringe movement only, which has by no means shown that the drive for more and better technology can or will be slowed. Nature still has claws, and so long as they give pain (no end is in sight), then technology will beckon people forward. The reality principle I am in search of has not been found—the concept of a limit to the possibilities of technology in providing happiness. Could it be found we would then know when to stop; and, even better, realistically know what to expect and hope for.

The other side of that coin relates to the social control of technology, and to the control of human beings to either make use of the fruits of technology or to neutralize its harmful effects. If a *biological reality principle* is needed to gauge the inherent limitations of technology, there is no less need of a *social reality principle* to set limits to the use of human beings by human beings for the sake of bowing to the logic of technology. To what extent should the undesirable consequences of technology be socially controlled, particularly when that means—the only meaning possible—controlling human beings? To what extent should the deliverances of technology be compulsorily applied? There can be no answer to those questions unless there is a sense of the limits of human malleability in social communities and of the requirements of the good society. Those requirements include a specification of the worth to be assigned to that which humans value, to that which they need as individuals and society, and to the tolerable costs of attempts to manage those human beings who are the agents of the uses and abuses of technology.

Even assuming that the necessary reality principles can be discovered, there would still remain the task of constructing from them a related set of ethical principles and decision-making procedures which would assure an effective respect for, and life in accordance with, the discerned reality. At that point, however, something more is needed than a new way of thinking or a better set of rational guidelines. Unless they could prove themselves capable of being lived, they could well end up in the historical junkyard of wise but ultimately meaningless reflections on the-way-life-ought-to-be. Freud objected to the "ethical standards of the cultural super-ego," which, he wrote, ". . . does not trouble enough about the mental constitution of human beings; it enjoins a command and never asks whether or not it is possible for them to obey it. It presumes, on the contrary, that man's ego is psychologically capable of anything that is required of it."[4] Could it be, for instance, that mankind must control technology, but that it will be psychologically impossible for it to do so?

The failure of attempts at world disarmament (despite the known and admitted hazards of an arms race dominated by nuclear weapons), the obvious reluctance of both the political powers and the political masses to take effective steps against environmental despoliation and pollution—to cite only some obvious examples—point to the existence of still another piece of reality. Human beings do not do that which they know they should do, and they do even less if the standards are raised too high. That observation is as old as Plato and as new as the daily deliberations of the United Nations. But it must be taken into account. If too much is required, prophets will flourish but few others. If too little is required, even less will be given. It may be that the population problem will eventually demand that attempts to control death be stopped. But it is unlikely—except by an imposition of the powerful upon the weak—that this demand will be heeded. And it may be that attempts to control the genetic quality of life will produce only social or biological disasters. But it is unlikely that the risks of these disasters will not be run, however foolish they might come to seem. The obvious problem here is that what people want is as much a reality to them and often as powerful in its force as the reality of that external world which frustrates those wants. The *psychological reality principle,* particularly with the introduction of advanced technology, is that people do hope to be happier by means of technology; that is an evident fact of contemporary life. If too great a renunciation of technology is demanded, there is likely to be none at all. A suitable and possible degree of technological asceticism must be determined.

I am deliberately attempting to define the problem which this book sets in the most difficult way possible. I have specified, on the one hand, the need for a technological reality principle, with both a biological and a social dimension, in order that the inherent limitations of technology may be discerned. But I have also specified, on the other, the limitations (which I believe to be inherent) in attempts to define and implement too perfect or high a set of ethical guidelines by means of which the reality principle can be respected. Human beings need to know that technology *cannot* deliver the total happiness they desire, the fulfillment of all possible hopes and dreams; that is already staring them in the face. But it is no less certain, from all historical evidence, that they will continue to hope for happiness through technology. Not even the worst imaginable nuclear or environmental disaster is likely to disabuse them of that hope. Are we then in the presence of an impossible, literally insane situation? Perhaps so, but it is the situation one must work with. It calls not only for a wisdom about what ought to be, but also and critically, about what actually can be; and "what can be" must, in the end, be something less than perfection. An ethical dialectic, therefore, between what should be and what can be is imperative in working through the problems of technology.

Freud spoke of the existence of a "cultural super-ego,"[5] which he contended has a profound influence on cultural evolution. As in the case of the individual super-ego, the cultural super-ego enjoins and commands. Even though Freud believed its demands are usually too great, at no point did he argue it could or would be dispensed with. This is a point of great importance. Beneath specific moral codes, and surrounding their application in practice, are a host of attitudes, biases and characteristic responses. At times, behavior will be no more than a reflection of them (mores). At other times, there will be a tension, often acute, between the actual mores and the rationally discerned good. One can define the moral life of a society as the way in which it works through the tension between what people actually do, what they believe they should do, and what they actually ought to do. This tension can be worked out in the direction of shaping moral standards to emergent practice, or in the direction of attempts to keep alive some transcendent ideals, even if it is recognized that these ideals are not being lived out nor are ever likely to be.

What remains, in any event, is the need for public standards of acceptable behavior, primarily defined in terms of limits and prohibitions. These standards can be set at a relatively undemanding level—the characteristic of societies which give individual freedom primacy; or at a high level—the characteristics of societies which give the common good or the general welfare primacy. Yet standards there must be, not only to regulate individual relationships, but also the relationship between individual and society. This is also to point up the necessary existence of a cultural super-ego, which will root those standards deeply in the psyche. Our problem is not only to define explicit cultural norms, but also to see if ways can be found to create the kind of cultural super-ego which technological societies require. That will mean a cultural super-ego which, in its commands and prohibitions, its aspirations and fears, is consistent with a biological, psychological and social reality principle.

Any attempt to describe ways of moving toward such an ambitious goal must proceed in a number of directions. There must be an acceptable definition of technological pathologies. What is to count as a pathological condition? There should be some understanding of the nature of technological thought and development, what I will call the "logic of technology." Insight is needed into the way technological developments and possibilities affect the relationship between the individual and the society. One must have some sense of what human beings actually need and, since there is often a difference, what they merely desire. Once some of these understandings are in hand, there will remain the normative task of proposing ways in which real and potential pathologies should be judged and controlled.

I do not believe there can be such a thing as value-free analysis, much less the logical absurdity of value-free prescriptions. The very choice of issues already shows one's hand. I want to argue that there are three key issues, the analysis of which will to a considerable extent dominate any understanding of the pathology of civilized, technological societies. The first is the meaning and significance of technology, the way it is understood and used. My contention will be that man is innately technological, and that it is impossible to understand the logic of technology without an understanding of man as a technological animal. The second issue turns on the conflict between individual and society, the tension between private goals and public need. The third issue is the relationship between private morality (individual or group) and public morality.

Of all these issues the one which I believe is at present

most vexing is that of the relationship between individual and society. I will use Sigmund Freud as a point of departure because I think he most accurately discerned some of the sources of the basic and abiding tension between the individual self and the surrounding community of other selves. Yet Freud needs correction, especially in gauging the impact of technology upon some old conflicts, as well as emendation, in order to move from his theory of culture to a valid social ethic. The most persuasive analyst of contemporary individualism is, to my mind, the sociologist Philip Rieff, who has attempted to show the ways in which science and technology have been utilized to enable the individual to negotiate more effectively (even if, as we shall see, not necessarily more validly) with a constraining society. Rieff's attempt to make use of Freudian insights will serve as the basis for some emendations of my own to the theses he has presented.

To bring my contention to a finer point, it can be said that the critical question to be decided upon is the way in which the fruits of technology are to be directed both to the good of the individual and to the good of the society. "Good," however, must be understood in several senses. It has almost always been the contention of those who give primacy to society that such a primacy is indeed most likely to guarantee the greatest good of individuals. The contrary contention has been that a focus upon individual good is most likely to result in societal good. Each of these opposing convictions has, in more sophisticated forms, presented itself not as one-sided but as the wisest way to achieve a proper balance between individual and social needs.

Christopher Lasch is essentially correct in his observation on the historical direction taken by science and technology:

> Modern rationalism revealed itself not only in the rational state and in the vision of a social order based on

> universal reason but in the unprecedented advance of science; and in a culture that set an increasingly high value on privacy, self-dependence, and personal fulfillment, it was perhaps inevitable that the achievements of modern science should be seen, not as a new stage in man's collective self-awareness, but principally as another means to personal fulfillment and the satisfaction of personal wants. The development of science and technology was part of the same individualistic rationalism that was already dissolving traditional concepts of social and political relations; and if at times the results of technology were dismaying to those who remained sensitive to the disruption of communal ties and the rape of nature, in the long run they seemed to justify themselves because they added to the sum of material goods, brought increased comfort, and above all seemed to enlarge the area of personal choice. The emerging concept of personal autonomy was thus bound up, not only with the idea of political and cultural freedom from institutional constraints, but with mastery over the biological conditions of existence.[6]

However accurate that analysis—with its clear implication that social rather than private needs should be the redirected goals of science and technology—it is a phenomenon of our day that the most thoroughgoing critique of an individualistic bias has emerged from a number of scientific quarters. Ecologists, zoologists, population geneticists and demographers have particularly taken the lead in pointing to the hazards of some forms of unfettered individual freedom: the freedom to procreate an excessive number of children, to ruin the environment and strip the earth of irreplaceable resources, to further pollute the gene pool by euthenics and the lack of procreative restrictions placed on the carriers of lethal genes. A considerable portion of this critique has used the survival of the species and the quality of species life as its focal point of value and morality. As I will try to argue, however, it is not at all clear that a morality of survival and species welfare offers the best possible antidote to an excessive individualism. Paul Ramsey has complained of a tendency in modern medicine to begin treating the desires of patients rather than their diseases. If that is a hazard, it may be no less a hazard to begin treating the diseases of the species rather than those of individuals.

Lasch has pointed to the evidence of a "new humility" on the part of medicine, a sense of uncertainty among scientists about whether everything they do or can do is for the good.

> Are we also beginning to understand [he asks] the limits of liberal individualism itself, which has furnished the political and cultural basis of the cult of technology? There cannot be a successful rebellion against the domination of social life by technology that does not incorporate a thoroughgoing criticism of liberalism . . . the inevitability of technological development has always been defended—at least in the West—on the grounds that it expanded the individual's freedom of personal choice. And until recently this was undeniably the case, at least for those able to avail themselves of the advantages of modern technology. It is only therefore when we find ourselves in private cars, marvellously mobile but unable to go anywhere because the highways are choked with traffic . . . it is only in short when we are confronted with the contradictions of individualism and private enterprise in their most immediate and unmistakable and by now familiar form, that we are persuaded—not persuaded but forced—to reconsider our exaltation of the individual over the life of the community, and to submit technological innovations to a question we have so far been careful not to ask: *"is this what we want?"*[7]

Lasch points effectively to one source of the pathology of technological communities, its individualism. But what are we to put in its place? To what degree should the community and even beyond that the whole human species be exalted over the life of the individual, an undeniable tendency in the new ecology and demography? There is no guarantee whatever that, even should such a new bias strengthen the possibility of human survival, such a survival would be worth the cost in social or psychological pathology. Enforced cleaning of the gene pool—by means of compulsory abortion of defective fetuses, mandatory sterilization of the carriers of hazardous genes, a refusal to treat or care for the victims of genetic disease (retinoblastoma and diabetes, for instance)—would go a long way toward doing away with liberal individualism. But it would do away with many other things as well, not all of which ought to be sacrificed to the communal well-being.

The concept of "the quality of life," on the whole a cheap, too-slick phrase, has at least the value of attempting to circumscribe some considerations which cannot be well managed by a simple concentration on the individual, on the one hand, or on the community or species, on the other. If some solid content is to be given to it, rescuing it from its too-easy identification with the cultural status quo, its content will have to include criteria for a critique of existing attitudes and desires, a respect for the biological, psychological and social reality principle and finally—with the greatest of difficulty—the outline of a viable theory of public morality. I do not see how society can take seriously the quality of life without some effective public morality, which finds ways of balancing private values and public good. The concept of "public morality," which though not the whole of the cultural super-ego is a critical part of it, runs, however, against the immediate obstacle in the West of a resistance to any suggestion of an enforced or even a democratically established consensus on normative standards. This is another expression of the dominance of individualism, which has exalted the right (and by presumption the capacity) of individuals to create their own world of meaning and their own ethical codes. It has been left to the legal and political system—with no intermediate, mediating social ethic—the task of managing the discrepancies between the myriad private moralities and the need for social peace and communal functioning.

Both population growth and the lure of control of genetic quality reveal the inadequacies of attempts to define public policy in the absence of a shared public morality, relying solely on legal processes. The family planning movement, which arose well before world population growth was recognized as a problem, was predicated on the putative right of individuals to control their own fertility without interference from society. The interference which that movement had in mind was primarily that of religious and cultural attitudes which, translated into legal and social prohibitions, effectively denied individuals the freedom of choice that would enable them to plan their family size effectively. But no sooner had this principle become accepted—at least as a principle—than an awareness of population problems forced some second thoughts. As Garrett Hardin, Kingsley Davis, Paul Ehrlich, Judith Blake and others have queried, how can there be any assurance that the supremacy of individual free choice will not lead, in its cumulative impact, to a rate of population growth harmful to social needs?[8] Why is there any reason to assume that the number of children a couple might quite freely want will turn out to be just the number that society can support? When Garrett Hardin, in the light of this question, asked for a "mutual coercion mutually agreed upon,"[9] he was, in his own terms, calling for what I would describe as a public morality, which in this case would establish a social norm and provide social sanctions against a procreative individualism looking only to the good of individual couples. And in realizing that the coercion would have to be "mutually agreed upon" he was, in my terms, recognizing that there would have to be a cultural super-ego which could sustain the coercion of individuals.

The same kind of tendency is evident in recent discussions of genetic quality. It is all well and good to refine genetic knowledge and a concomitant medical technology to the point where couples can have the kind of healthy, genetically sound children they want (or are persuaded they should want). That has been the traditional goal of genetic counseling, and one more reflection of the dominance of individualism in medicine as well as the rest of science. But does not society have some rights in determining for itself the genetic quality most suitable for its common needs? Why is there any reason to assume that the kind of children which couples want represent genetically the kind of children who are good for the species? If "mutual coercion mutually agreed upon" is an appropriate goal to assure a control of population quantity, why is it any less appropriate to assure population quality? The latent demand, then, for a public morality of genetic quality (with appropriate legal and social sanctions to back it up) is no less apparent in some branches of the genetic guild than is a public morality of population quantity in the population guild.

From a variety of sources, then, liberal individualism is under attack. If there is any common thread underlying the various antidotes which have been proposed, it is the need for fresh communal ethical standards which will not only allow new community-oriented laws and policies to be enacted, but will sink deeply enough into the cultural super-ego that the new laws and policies will seem a long-overdue opening to a new balance and sanity. While it will become clear that I am sympathetic with this direction, I believe it is by no means clear how a public morality is to be constructed, one which would rest on the proper reality principles but would not, in the process, simply serve as one more disguised version of tyranny.

To the need for a science of limits for technology—biological, social and psychological—must also be added, then, a solution to the problem of the individual and society. A first step in this process is to look at the meaning and logic of technology, focusing in particular on two related ways of understanding the significance of technology: on the one hand, the place of technology in ensuring and enhancing survival and, on the other, its place in fulfilling human desires which go beyond survival. My aim will be to show that an excessive orientation toward group survival, with technology as the chosen means, is as damaging as an excessive emphasis upon the personal "liberation" possible through technology. I can do no better, however, than to turn to Freud for a fundamental understanding of the human condition which provides the setting for the problems of technology that I want to consider.

## NOTES

1. Sigmund Freud, *Civilization and Its Discontents,* trans. Joan Riviere (Garden City, N.Y.: Doubleday, 1958), p. 104.
2. *Ibid.*
3. *Ibid.*
4. *Ibid.,* p. 102.
5. *Ibid.,* p. 104.
6. Christopher Lasch, "Birth, Death and Technology: The Limits of Cultural Laissez-Faire," *Hastings Center Report,* Vol. 2 (June 1972), pp. 1–2.
7. *Ibid.,* p. 4.

8. See Garrett Hardin, "The Tragedy of the Commons," *Science,* Vol., 162 (December 13, 1968), pp. 1243–48; Kingsley Davis, "Population Policy: Will Current Programs Succeed?" *Science,* Vol. 158 (November 10, 1967), pp. 730–39; Paul R. Ehrlich, *The Population Bomb* (New York: Ballantine Books, 1968); Judith Blake, "Population Policy for Americans: Is the Government Being Misled?" *Science,* Vol. 164 (May 2, 1969), pp. 522–29.
9. Hardin, *ibid.,* p. 1247.

# CHAPTER 2

# *Pathologies of Civilized Communities*

IT IS NOW TRIVIAL to say that Western culture is undergoing a crisis, but it is not trivial to live it. To live it and not just talk about it means that one takes upon one's shoulders, willingly or unwillingly, all the burdens of confusion, uncertainty and a clouded vision. No human being can for long enjoy living with a weight of that kind. Up to a point he can celebrate change and look warmly toward a different future; change can be enjoyable and a challenge—so much the rhetoric of moon walks, positive thinking and the deification of technology keeps telling us. But as often as not it presents itself as a dagger, cutting our stabilities to pieces and hacking away at our carefully constructed inner worlds. There are, to be sure, those who prosper from social flux: the advertisers, money lenders, arms and space agencies, builders and road pavers. They live off consumption, taste, quantity and tension; more is better, and change welcomed. Producing the hardware and the software of social and technological flux is profitable. But there are also those who do not prosper.

As Freud saw the problem of civilized life, technology only exacerbated a very old problem: How are human beings to live together? Technology, particularly in its refined instruments of war, means that people now have available more efficient methods of killing each other. But the reasons why they want to kill each other predate those instruments, going back to the nature of man, both historically in time and psychologically in inner space. Even so, however, Freud himself leaves open the possibility that, if nothing really new has been introduced into the society of man, the very fact that society is now more advanced than ever, more "civilized," means that the power of destruction, aggression and death have been intensified. Yet it was the root causes of human aggression which intrigued Freud; a "pathology of civilized communities" would have to take them into account. Then, and only then, could one begin trying to understand how the particular qualities of contemporary life heightened and flavored some very old struggles.

A good place to begin is with Freud's own formulation of the problem, using as our main text *Civilization and Its Discontents,* that small gem of a book in which, toward the end of his life, he spoke so eloquently yet so simply to the matter of preserving civilized life. In *The Future of an Illusion,* written two years before (1927), Freud had pressed the great need for "reconciling men to civilization."[1] In his critique of religion, he had argued the necessity for turning man's attention to the earth; religion, which turned man's eyes, so he believed, toward another world, was a hindrance to the task of reconciliation. "Men cannot remain children forever; they must in the end go out into 'hostile life.' We may call this *'education to reality.'* "[2] It is this larger task, the education of men to reality, which preoccupies Freud in *Civilization and Its Discontents.* Clearing away the "illusion" of religion is a first step; many more remain to be taken, for what is at stake is the survival of mankind, now cursed with the technical means to exterminate itself.

The question which preoccupied Freud stems from a paradox. Why is it that man, who depends upon civilization for his survival, is so hostile to civilization? Human beings cannot do without the company and the help of their fellows, and yet they cannot abide it either. There is a fearful mystery here, not a matter of idle speculation:

> The fateful question of the human species seems to me to be whether and to what extent the cultural process developed in it will succeed in mastering the derangements of communal life caused by the human instinct of aggression and self-destruction. In this connection, perhaps the phase through which we are at this moment passing deserves special interest. Men have brought their power of subduing the forces of nature to such a pitch that by using them they could now very easily exterminate one another to the last man. They know this—hence arises a great part of their current unrest, their dejection, their mood of apprehension.[8]

If anything, these words have a greater pertinence now than when they were written. Technology has made massive strides since the 1920s when Freud wrote that passage. The unrest, the dejection, the apprehension which Freud saw even then have increased. The "derangements of communal life" are more evident than ever, in the revolt of youth, in a turning away from public to private life, in the decay of the cities, in rioting, in a seemingly universal alienation. If the traditional religions of the West show signs of decline, religiosity is stronger than ever, expressive of a desperation which, whether in astrology, drugs, sensitivity training, or ideological fanaticism, grows apace. How did we get here?

Freud tries to answer this question for his own time, not by an examination of the idiosyncrasies of modern culture, but by asking us to look at human behavior in its most ordinary manifestations. If we want to know what men take "the purpose of life" to be, then we can make a start by seeing what they do. The answer, he believes, is evident enough. Men "seek happiness, they want to become happy and remain so."[4] Negatively, they aim to avoid pain and discomfort; positively, they want to experience intense pleasure. The "pleasure-principle" is in control here. From the moment of birth, it sets forth a plan: pleasure is to be sought and increased, pain is to be avoided. But it is a plan which shows no promise of full realization, for the scheme it sets forth "is in conflict with the whole world, with the macrocosm as much as with the microcosm. It simply cannot be put into execution, the whole constitution of things runs counter to it; one might say the intention that man should be 'happy' is not included in the scheme of 'Creation.' "[5] Freud's initial pessimism rivals that of Ecclesiastes. Even when we achieve pleasure, the experience is transitory, unable to sustain itself, ever dependent upon the recurrence of intense contrasts between pent-up needs and their temporary satisfaction.

The limitations of our bodies pose the first obstacle to the achievement of enduring pleasure. We suffer from our body, which is born to decay and die; the sources of pleasure are more limited than the sources of suffering. We suffer from nature, whose arsenal is full of weapons of destruction. We suffer from our relations to other men, a pain all the more difficult to bear because it seems so unnecessary: we have enough problems already, living with our finite bodies in the midst of the muscularities of nature.

But of course man fights back, even if he soon learns that it is wise to reduce his demands for happiness. There are any number of possibilities available, which Freud analyzes one by one. He can seek a total gratification of all his desires—if he is willing to throw caution to the winds and pay the penalty attached to self-indulgence. He can try to influence his body chemically to produce a maximum of pleasure, by drink, by drugs or, now, by electrical manipulation—if he is willing to be dependent upon these things, which can in many circumstances harm him and which can also be taken from him. He can attempt to annihilate his instincts—if he is willing to settle for peace alone and to give up all other activities. He can attempt the way of love—if he is willing to tolerate the frustrations and disappointments this will bring. He can seek to sublimate his instinctual life in intellectual or artistic work—if he has special gifts and talents, and if he recognizes that in the end his body will fail him and that sublimation offers no sure protection against the forces of fate. He can be a religious believer—if he is willing to live in a world of illusion and if he is willing to see his energies diverted from a coping with reality. Freud does not say that any of those options are wholly unsatisfactory. He only says that none of them can be had without paying a price; something else must be excised to make the option viable. And none of them is proof against the body, nature or other men.

Unfortunately, the wisdom of pessimism and reduced demands for happiness advances us only so far. We cannot in a stoic restraint let go of the drive for happiness: "The goal toward which the pleasure-principle impels us—of becoming happy—is not attainable; yet we may not—nay, cannot—give up the effort to come nearer the realization of it by some means or other."[6] Whether we choose, according to our disposition, a negative or a positive path, the avoidance of pain or the pursuit of pleasure—and choose we must—we will not succeed; we cannot have everything we want and we will not. At the least, caution is in order. If we cannot remove ourselves from the playing field, we can play the game warily, not trusting in happiness from one quarter alone and hoping we have the good fortune to be born with a mental constitution able to adapt itself to the world.

But has enough been said? The advice to be cautious, not to put all of our happiness eggs in one basket, to reduce our demands, is banal and Freud knows it. Would that the problem could be dealt with so easily. It is possible, Freud says, for human beings to recognize and finally accept that suffering which comes from the body; we can see the inevitability there and submit to it. As for the abolition of the suffering which nature imposes upon us, whether from within the body or from the outside, there are good grounds—think only of the advances of science and medicine—for hoping for a mitigation of our pains for a time; and that is something.

But why must we suffer from other men? That suffering seems gratuitous. There appears no good reason why human beings should kill each other, why they cannot live in peace, why they cannot desist from all those violent forms of aggression whose kind and variety seem innumerable. But this is the reality of human community. That very set of human relationships which proves so successful in the common fight against nature carries its own seeds of destruction with it. Is nature to be conquered only at the cost of mutual human destruction? What then is the gain? We can readily enough try to explain the derangements of human community by an examination of the defects of the economic and political arrangements existing in various states and nations; prescriptions for reform abound. But it has always been so and, while the possibility of progress exists here, the unceasing recurrence of the derangement throughout history suggests that something deeper is at stake, "that a bit of unconquerable nature lurks concealed behind this difficulty as well—in the shape of our own mental constitution."[7]

There is every sign that Freud recognized how unpleasant this suggestion would seem. To locate the source of human unhappiness in human nature itself, to see in a sense that the drive for community must suffer the weight of an "unconquerable" given, to say that the problem goes beyond the manipulation of political and social systems, is to introduce a profound pessimism, the very death of utopian visions and unlimited hope. It means that the one realm of nature seemingly open to a full human conquest—the life of human beings together—may also be closed, fated to as bad and as determined an end as our private bodies, whose organs will inexorably decay and bring us down. The Christian doctrine of original sin is a way of expressing the insight that the trouble lies in man himself, something he and his kind are born with. And the doctrine of the redemption—where one man dies for the sin of all, thus releasing all from their bondage—offers a vision of a final salvation. Freud will not let us off so easily. Final redemption is but a wish and there is nothing in reality to support a happy outcome for that wish. There is open to us only the harder way of understanding—a naked confrontation with the "reality principle"—and the piecemeal business of taking small steps forward.

Before we can move forward, though, we must understand. Why is it, one must first ask, that so many people are hostile to civilization? That question comes first because the fact of human aggression immediately presents itself to our eye. Having created a civilization to ward off nature, that human life may survive, the creators immediately set about harming each other. As if their struggles with nature were not enough, they seem bent on creating struggles with each other; nature itself is hardly so ingenious as man in throwing up obstacles to happiness. Where the reality principle first forces the individual child to take account of the external world, which inflicts pain as the outcome of an unbridled attempt to live out the pleasure principle, so too humanity as a whole is forced to take account of the reality of human community, which inflicts pain on a massive scale. If man must be protected from nature, he must also be protected from his own kind. In its rough outlines, Freud's model is a simple one: each human being seeks his own happiness, independently of others; but civilization requires the cooperation of all and that means the renunciation, to an indeterminate but considerable degree, of private happiness. The strivings of the individual's pleasure drive must be curbed; the good of the whole takes precedence, for only the whole, together, have the power to curb the ravages of nature.

We come quickly to the point:

> Thus the cry for freedom is directed either against particular forms or demands of culture or else against culture itself. It does not seem as if man could be brought by any sort of influence to change his nature into that of the ants; he will always, one imagines, defend his claim to individual freedom against the will of the multitude. A great part of the struggles of mankind centers around the single task of finding some expedient (i.e. satisfying) solution between these individual claims and those of the civilized community; it is one of the problems of man's fate whether this solution can be arrived at in some particular form of culture or whether the conflict will prove irreconcilable.[8]

I have tried so far to let Freud speak for himself and to analyze his meaning. But some critical objections are called for in order not to concede any more to Freud's harsh rendering of the human situation than is necessary. For Freud does, in his thinking, lay an almost inescapable snare. On the one hand, he seems to leave little room for a positive vision of culture, one which would recognize that long and valuable intellectual tradition which has seen man as a social and political being. And on the other hand, he appears to preclude the development of a social ethic which would consist of anything but a necessary attempt on the part of civilization to keep its pleasure-seeking, self-centered individuals under control.

On the first point, it is well to remind ourselves that the experience of human community, which sees the individual working with others for the good of the whole, can be a joyful, satisfying one. Human beings do not gather together just for parallel play, or for the sake of mutual protection against nature. Even psychologically, most find isolation difficult to bear; they are "not themselves" when cut off from all human contacts. More than that, the very phenomenon throughout history of utopian visions of a better life together contains more elements than simply a satisfaction of the instinctual life of individuals. They also contain visions of communal happiness, of a common art, politics, culture and sensuous pleasure; they envision as well a sharing not just of the burdens of life but also of its joys and satisfactions. There is no reason why a recognition of the historically multitudinous concepts of society as a positive, mutually enriching human experience cannot be encompassed within the reality principle also. That recognition does not preclude an equally sharp perception of the constant conflict between individual and community; it only complements it by including a wider range of reality than Freud appears willing to allow.

On the second point, it is exceedingly difficult to see how Freud's analysis of the tension between individual and civilization can lead to anything other than an ethic of social repression with survival as the dominating value. By casting the individual and society as antagonists, the ground is laid for legitimating the policeman as the final symbol of cultural unity and preservation; "law and order" quite properly becomes the dominating slogan and political fact of life, with "freedom" the enemy as soon as it endangers good order and "justice" the weapon with which each is repressed in the name of all. Again, though, there are other visions—which bespeak a wider reality than Freud envisioned—that see the possibility of a social ethic where law and order become only the minimal requirement of a society which seeks in its common life to transcend the need for a common control of all by all. Freedom, then, is not seen simply as the expression of the individual against a repressive society, but as a means whereby the individual finds the most appropriate way to join his instincts, talents and inclinations to a common community which thrives on the variegated contributions of its different members. And justice, far from being a tool of repression or a mere principle of distribution of goods, rights and obligations, is seen as a way of conceptualizing the importance of individual dignity for the common business of living together and mutually constructing a society in which the value of the whole can be greater than that of its individual parts.

While Freud was wary of drawing overly close parallels between the development of the individual and the development of culture, he saw similarities which were nonetheless striking. The first lesson a child must learn is that there is a world out there, a world which will not always let him do what he wants. If the child, seeking pleasure, has his demands, reality has some as well; yet the struggle is an uneven one, and the individual has to come to terms as much with himself as with the reality outside him. The instincts which clamor for unlimited gratification—a hopeless venture—must be redirected along paths that promise some degree of possible satisfaction.

Freud frequently employed the language of economics to describe the psychological transactions which must take place: resources are to be preserved, trades must be arranged, a balance of payments should be achieved. Sublimation constitutes the psychological process whereby the aims of the instincts are realigned with the actual possibilities of gratification afforded by reality. But this does not preclude a good bit of shrewd bargaining. If the upper limits of the bargaining process appear fixed—one can't expect that the balance of payments will always be in one's favor (death finally terminates all transactions in favor of nature), there is considerable room for maneuver. Fortunate indeed is the person born with the right temperament at the right time and in the right place to be a successful entrepreneur of the instincts. Yet every person, fortunate or not, will have to bargain. Under the best of circumstances life presents a harsh economy. What is the purpose of life? Freud's answer seems to be that it is to manage, if one is gifted, to transact some shrewd bargains; and if one is not gifted, then at least to keep oneself from being cheated by life at every turn.

The complication is that everyone else is bargaining at the same time. The developing child must learn not only how to bargain with a demanding nature, and with the limitations of his own body, he has also to do some trading with others as well. Not only are they seeking to gratify their own pleasure principle, they are also all attempting, as a community, simply to survive. That there are not enough goods to go around, whether money, or food, or women, means that scarcity limits the possible success of individual strivings. The minimal price to be paid just to make use of the goods which are available is that humans must band together, and that means the individual will have to suppress his instincts as the admission fee to life in the world of other human beings. The only way one can have the slightest chance of satisfying the pleasure drive is to survive, and one can survive only by joining with the rest of humanity to overcome nature. That is, inexorably, to be forced to start off on the wrong foot so far as the demanding pleasure principle is concerned. A fine dilemma, that. "Sublimation of instinct," Freud wrote, "is an especially conspicuous feature of cultural evolution; this it is that makes it possible for the higher mental operations, scientific, artistic, ideological activities, to play such an important part in civilized life. If one were to yield to a first impression, one would be tempted to say that sublimation is a fate which has been forced upon instincts by culture alone. . . . it is impossible to ignore the extent to which civilization is built up on renunciation of instinctual gratifications."[9]

But why, one must ask, does Freud feel compelled to classify science, art, politics and all higher mental activities as something forced upon the instincts by culture alone? Freud is most unsatisfactory on this point, obsessively reductionistic at just that point where a concern for the reality principle would seem to demand a wider understanding of an important range of human characteristics. At the very least, Freud might have been willing to see that the scientific enterprise, which he extols, must represent something more than mere sublimation; it has sought more than can be explained by an attempt to deal with repression of instinct, unless of course one cares to say that the work of Newton, Darwin, Einstein and Freud himself represents only rich variants on a life of culturally enforced sublimation. That civilized life makes the higher mental operations possible ought to have said more to Freud than it did. It is just as plausible that, for reasons quite other than sublimation, the higher mental operations are pursued because they offer individual and cultural satisfactions quite as attractive and compelling as the fulfillment of instinctual demands. Put in more narrow terms, Freud has offered a theory which is unverifiable in principle, for even an attempt to verify it (requiring higher mental operations) could be taken as an illustration of his point.

From the viewpoint of the need to create a social ethic, a still more fundamental criticism is possible. It is perfectly conceivable that a social ethic could be built solely upon the principle that the meaning of civilization is repression and nothing else; thus a civilization of law and order is the highest common good. But a richer social ethic could be built upon a base which construed art, politics, science and religion as human attempts to give life a meaning which goes beyond satisfying only the instincts of individuals. In point of fact, it is doubtful that humans will for long accept a social ethic which does not, in its foundations, give a place to the human desire for a structure of meaning and value which combines the good of the individual with the good of the community, with each enriched by the other. That has been the history of ethics.

While the conditions of civilized life require sublimation—scarcity alone is enough to show its necessity—an even more ominous reality can be discerned. Tentatively in theory, yet firmly in conviction, Freud believed that a desire for aggression "has to be reckoned as part of man's instinctual endowment." The record of history, if one needs more evidence than that to be had by opening one's eyes, is there for all to see: murder, war, rape, aggression, lying, stealing, cheating—the list is long and historically stretches back to the beginning of human life. Knowing that, and knowing too that nature must be subdued, it is necessary that men impose restraints upon each other. What people will not do voluntarily they must be made to do. Freud's comment on Communism can stand as a model of his doubts that social or political reform could end human unhappiness, that somehow it is the way in which society is organized that is at the root of human aggression: "I have no concern with any economic criticisms of the communistic system. . . . But I am able to recognize that psychologically it is founded on an untenable illusion. By abolishing private property one deprives the human love of aggression of one of its instruments, a strong one undoubtedly, but assuredly not the strongest. It in no way alters the individual differences in power and influence which are turned by aggressiveness to its own use, nor does it change the nature of the instinct in any way."[10] Neither the abolition of private property, then, nor the removal of all sexual restraints would do away with aggression. There is no reform which could be imagined, however radical, but "that the ineffaceable feature of human nature would follow wherever it led."[11]

Some societies may require more sacrifice of individual desires than others, and some reforms may ameliorate the inequities and injustice which spring from a maldistribution of power and freedom. But all cultures will require some serious degree of individual sacrifice. If men are to be protected from nature and from each other, the first requirement is that they give up their freedom. Left open is *how much* freedom they must relinquish. Knowing that they are limited, from the inside as well as from the outside, it is hardly surprising that humans should not be happy. They start their life in the midst of a bad bargain. "If civilization requires such sacrifices, not only of sexuality but also of the aggressive tendencies in mankind, we can better understand why it should be so hard for men to feel happy in it."[12] If the human race can be divided among those willing to entertain visions of a utopian society and those whose sense of realism forces them to settle for something less, then Freud must be included among the latter. When Freud says, on the one hand, that civilization may be made more satisfactory—for he by no means denies us all hope—but says also, on the other, "that there are certain difficulties inherent in the very nature of culture which will not yield to any attempts at reform,"[13] we know we are in the presence of one who will not allow, and would not have us allow in ourselves, our hopes to run ahead of our good sense. And as our good sense made clear to us as children, there is a stubborn reality out there in the form of nature and other people, and a reality within, in the form of our own body; they cannot be wished away.

Yet Freud's own understanding of the private and public reality of life led him to press further. Granted a conflict between the individual and culture—*any* individual and *any* culture—the mystery is hardly dissolved; its outline only has been described. Why does there exist in man a tendency toward aggression, a tendency so inimical to human happiness? Freud had long puzzled over this question and in *Beyond the Pleasure Principle* (1920) he had put forward the hypothesis, one of his boldest, that there must exist in man a death instinct, of which aggression is a primary manifestation. Freud was led to this theory through his attempt to understand the meaning and purpose of compulsive repetition. Not only children but even adults show a tendency to repeat those acts or fantasies which bring them pleasure; clearly this repetition is a source of pleasure in itself. In searching for the dynamic behind this compulsion, Freud theorized that some attribute of all instinctual life might be at issue and, beyond that, perhaps a fundamental characteristic of organic life in general. That attribute, Freud guessed, might be an inherent urge in all things organic to return to an earlier state, to turn backward and restore a state of stability once possessed but lost because of external pressures.[14]

> Let us suppose, then, [Freud wrote] that all organic instincts are conservative, are acquired historically and tend toward the restoration of an earlier state of things. It follows that the phenomena of organic development must be attributed to external disturbing and diverting influences. The elementary living entity would from its very beginning have had no wish to change; if conditions remained the same, it would do no more than constantly repeat the same course of life. . . . If we are to take it as

> a truth that knows no exception that everything living dies for *internal* reasons—becomes inorganic once again—then we shall be compelled to say that *"the aim of all life is death"* and, looking backwards, that *"inanimate things existed before living ones."*[15]

The essence of the death instinct, therefore, is the thrust toward return to an earlier organic state; it is in this sense that Freud can refer to the death instinct as essentially "conservative."

Yet it is also apparent that there is a life instinct as well, for life tries also to preserve itself; there is an internal impulse toward progress and more advanced development.[16] This impulse can be called conservative as well; both the life and the death instincts seek to preserve some fundamental state, the former that state of life already existing, the latter that state of nonlife which once existed. At the same time, however, Freud hypothesized that the life instinct, Eros, seeks not only higher stages of development but also an ever-widening scope of unity. The life instinct seeks both self-preservation and union with other organic substances. But if the life instinct moves in one direction, and the death instinct, Thanatos, in another, then tension and interaction between the two is inevitable: ". . . beside the instinct preserving the organic substance and binding it into ever larger units, there must exist another in antithesis to this, which would seek to dissolve these units and reinstate their antecedent inorganic state; that is to say, a death instinct as well as Eros; the phenomena of life would then be explicable from the interplay of the two and the counteracting effects on each other."[17]

The problem Freud faced at this point was in explaining the fact of aggression and, in particular, the relationship of Eros, seeking to preserve life, to that of aggression. His solution was ingenious. Could it possibly be the case that Eros, recognizing that the death instinct works against it, turns its force away by directing it to the outer world? This struck Freud as a plausible possibility:

> The idea that part of the instinct [Thanatos] became directed towards the outer world and then showed itself as an instinct of aggression and destruction carried us a step further. The instinct would thus itself have been pressed into the service of Eros, in that the organism would be destroying something animate or inanimate outside itself instead of itself. Conversely, any cessation of this flow outwards must have the effect of intensifying the self-destruction which in any case would always be going on within. From this example one could then surmise that the two kinds of instincts seldom—perhaps never—appear in isolation, but always mingle with each other in different, very varying proportions, and so make themselves unrecognizable to us.[18]

If the mingling that Freud speaks of is most visibly evident in sadism, it is no less likely to be present in the drive of the ego, giving vent to its omnipotence wishes, to master nature. If man *must* conquer nature in order to survive, he also *wants* to conquer nature as an expression of his omnipotence; and this conquest will take the form (though often benign) of an assault upon nature—we need houses to live in but there is also the destructive need, quite apart from the goal of shelter, to kill the trees which produce wood for us. If we carry Freud's thinking into the present ecological crisis—marked by air and water pollution, the ruination of forests and the misuse of natural resources—something more than sheer bungling and mismanagement would seem at issue; mankind may be acting out its aggressive impulses. The same may be true of efforts to "conquer" disease.

Yet once one tries to carry the Freudian image of man's

relationship to nature as far as it can go, problems appear. While human beings have destroyed and ravished nature, they have also cultivated and enriched it. They have developed gardens, whose aim is beauty, and they have cleared and cultivated otherwise useless scrub and swampland. Wood has been used not just to build houses, but also to create statues, masks and other artifacts; something is done with a part of nature which is badly and misleadingly described as simply one more instance of aggression and destruction, or even as an expression of the need to survive the hostilities of nature. A viable competitive conception of man's relationship to nature is that of mutual interdependence and mutually enriching possibility; nature can be improved by man while at the same time remaining faithful to nature.

By means of developing one hypothesis after another—and that is Freud's resolutely nondogmatic method here—the central problem of culture comes into focus. The work of Eros, as it seeks to expand its scope, "aims at binding together single human individuals, then families, then tribes, races, nations, into one great unity, that of humanity."[19] But working against this unity is the instinct of aggression, where each human is hostile to all others. Civilization, dependent for its survival and progress on the unity brought about by Eros, is thus thwarted; or better, it is opposed, and the two instincts struggle with each other for the upper hand. "And now," Freud is able to conclude, "the meaning of the evolution of culture is no longer a riddle to us. It must present to us the struggle between Eros and Death, between the instincts of life and the instincts of destruction, as it works itself out in the human species. This struggle is what all life essentially consists of and so the evolution of civilization may be simply described as the struggle of the human species for existence."[20]

Yet if Freud has, in his view, set us on the path to an understanding of the perennial problem of civilized life—and one can now see why he believed the inherent difficulties of culture go beyond those wholly subject to social reform—there remains the problem of understanding the ways in which civilization attempts to control the instinct of aggression. It is here that we are provided an insight into another of Freud's more pessimistic observations: the greater the degree of civilized life, the greater the intensity of aggression. The more obvious ways in which civilization controls aggression are well known: laws, mores, punishments, imprisonment. But Freud felt that more than these externally coercive measures were required; by themselves they would be insufficient. It is what goes on in the developing individual which is decisive. Born with an instinct to aggression, a child somehow grows up with a set of built-in curbs. Instead of projecting all of the aggression outward, most of it is turned within. The ego, which would if it could vent its aggressiveness, is controlled in its expression by the super-ego, which in the form of conscience exercises an aggressiveness toward the ego itself. The result is guilt, the tension between the demands of the ego and the internalized punishments of the super-ego. The work of punishment and control, originally undertaken by external authorities, is taken over by the internal authority of the super-ego. Thus is culture aided by the workings of the mind, which produces guilt, a useful commodity to civilization which could not, even if it wanted to, set a policeman next to each individual (and how would one control all the policemen?). In the dread of authority, during the earliest stages of individual development, and in the dread of the super-ego, in the later stages, are to be found the sources of guilt.

If all of this is highly useful for a civilization which needs to keep its people in check, it is a psychological process with some highly dangerous by-products. If Freud's implicit motto seems to be "nothing good can be had without bringing about something bad," it finds acute expression in this instance, when one sees what is produced by the internalization of authority and ensuing guilt. While external authority may be appeased by obedience or, if a crime is committed, satisfied once punishment has been meted out, the super-ego is a much more demanding master. While the renunciation of a behav ioral manifestation of instinctual demands will satisfy the authorities of the world (though Freud may have been too optimistic here about the nature of modern totalitarian regimes), it will not satisfy the super-ego, which will allow neither the behavior nor the wish which would, if it could, manifest the behavior.

The gratification renounced, the wish remains—and the super-ego knows this, knows it and does not forgive it: "Renunciation no longer has a completely absolving effect; virtuous restraint is no longer rewarded by the assurance of love; a threatened external happiness—loss of love and punishment meted out by external authority—has been exchanged for a lasting inner unhappiness, the tension of a sense of guilt."[21] Moreover, renunciation and guilt stand in a dynamic relationship to each other: if it is conscience which originally gives rise to renunciation, soon the renunciation itself begins generating further pressure from conscience, which in turn demands still more renunciation. Thereby is set in motion a deadly escalation: the more that is given, the more that is demanded; guilt keeps increasing and no end seems in sight.

However hard this escalation is on individuals, it also turns out to have an effect upon civilization as well. The drive of Eros toward ever-encompassing and enlarging circles of human unity, a prime manifestation of civilization, requires that the same civilization increase its guard against the aggressiveness which would destroy the unity; life must be kept moving forward, resisting the power of death to draw it back to nonexistence. But this can be done only by increasing the sum of guilt. The greater the degree of unity and thus of civilization, the greater the need to suppress aggression; and the more aggression is suppressed, the greater the degree of individual guilt which must be borne. "If civilization is an inevitable course of development from the group of the family to the group of humanity as a whole, then an intensification of the sense of guilt—resulting from the innate conflict of ambivalence, from the eternal struggle between the love and death trends—will be inextricably bound up with it, until perhaps the sense of guilt may swell to a magnitude that individuals can hardly support."[22] Thus guilt, ever intensified by the renunciations demanded by a civilized life in common, grows apace with progress.

If civilization is to advance, which Eros drives it to do—and Freud instantly rejects the romantic notion that we can return to the kind of life lived by primitives (who had, in any case, no real freedom as individuals)—then the happiness of individuals must be sacrificed; it is the upward spiral of guilt which ensures this. The argument of *Totem and Taboo* (1913), which postulated a primal horde killing a primal father, a father both loved and hated, was that the death of the father brought a primal sense of guilt. With this guilt culture began, for the authority which the father by his power had imposed from the outside was now imposed from the inside, first by the primal horde upon itself and then by the process of internalization within each individual. Guilt was and remains the key. Just as it is guilt which keeps the instinctual life of the individual in check, so too civilization itself is dependent upon the pervasiveness of guilt to ensure its protection from its own members. The wished-for aggressions remain uncommitted; but they are still wished-for and punished by remorse.

It is striking how often Freud feels compelled to repeat and restate his argument in *Civilization and Its Discontents.* Each time he apologizes for doing so, but evidently feels that his line of thought can easily be misunderstood. One can see why. While the individual components of his argument are clear enough, their interrelationship becomes increasingly complex. Freud was attempting nothing less than an explanation of the discontents of civilization which took into account the psychological development of the individual, the evolution of culture, and the nature of all organic life—a large order for such a small book, however much the ground may have been prepared in Freud's earlier writings. But he saw the pressing need, enough to daunt most men, for a theory of culture which was at once cosmology, biology, anthropology and psychology. Beginning with a cosmology which envisioned nature as constantly seeking equilibrium, and with an anthropology which imagined an original primal horde driven to kill its father, and with a psychology which saw both the life and the death instincts as carrying out the organic demand for conservation, he was finally led to a field-theory of culture. One may not like the fact that this field-theory postulates the inexorability of discontent in civilization, but Freud felt no other possibility could take full account of all the evidence.

The parallels between the development of the individual and the development of culture are, Freud believed, too unmistakable to be set aside: the individual aims to be incorporated into the group (he has no choice here; he cannot survive without the group nor can his instincts be gratified); and the group aims to incorporate the individual into it (it has no choice either; it needs him for its own survival, in a positive sense, and it cannot afford not to pacify him, in a negative sense). Yet there is an important difference also between the aims of the individual and the aims of the culture; and in that difference lies the most vexing tension. Freud saw in individual development

> the interplay of two trends, the striving for happiness, generally called "egoistic," and the impulse toward merging with others in the community, which we call "altruistic." . . . In individual development . . . the main accent falls on the egoistic trend, the striving for happiness; while the other trend, which may be called the "cultural" one, usually contents itself with instituting restrictions. But things are different in the development of culture: here, far the most important aim is that of creating a single unity out of individual men and women, while the objective of happiness, though still present, is pushed into the background; it almost seems as if humanity could be most successfully united into one great whole if there were no need to trouble about the happiness of individuals.[23]

At this point, however, having said the worst—for what, in a sense, could be worse than the dark possibility that the happiness, peace and harmony of humanity required, finally, ignoring the need for individual happiness and freedom?—Freud opens the door just a crack. It is not likely, he says, that the antagonism between Eros and Thanatos can be overcome. They are part of the nature of organic life, and thus probably beyond human control. But the antagonism between the individual and society may not be fated to so poor a prognosis. To be sure, the individual's drive toward happiness—on his own terms, which are unlimited and resistant to any restrictions—will have to struggle with his drive for unity with the rest of humanity. But that is a different struggle from that between Eros and Thanatos: "it is a dissension in the camp of the libido itself, comparable to the contest between the ego and its objects for a share of the libido; and it does eventually admit of a solution in the individual, as we may hope it will also do in the future of civilization—however greatly it may oppress the lives of individuals at the present time."[24]

This seems to be Freud's way of saying that the dual striving that he called "egoistic" and "altruistic," both of which are the product of individual development, admit of some degree of reconciliation. But he provides no clue about how this reconciliation might be effected, nor does he say how it could serve to overcome the guilt and resultant increase in aggression which the development of culture imposes. In the last sentence of *Civilization and Its Discontents,* Freud says, "And now it may be expected that the other of the two 'heavenly forces,' eternal Eros, will put forth his strength so as to maintain himself alongside of his equally immortal adversary."[25] Again, though, Freud provides no reason why this expectation should be entertained: and it is an expectation, after all, which only hopes that the strength of Eros will be at least equal to—and no more than that—the strength of Thanatos. It is a very tiny candle which Freud lights, seemingly despite himself. At the end of the penultimate paragraph in *Civilization and Its Discontents* he had said: "My courage fails me, therefore, at the thought of rising up as a prophet before my fellow-men, and I bow to their reproach that I have no consolation to offer them; for at bottom this is what they all demand—the frenzied revolutionary as passionately as the most pious believer."[26] It is as if, having said that, Freud could not quite bring himself to leave his readers utterly bereft; some consolation, however small, had to be offered. Freud the man, perhaps, was willing to give what Freud the theoretician could not. In his own terms, though, the hope Freud offers appears insupportable.

Though Freud provided few clues about why mankind might hope for a surge from the power of Eros, and in that respect provided no metaphysical grounding for the small hope he held forth, it is apparent that he believed man did have one tool to make the best of things, and even to make things better. As an heir to the enlightenment, and as one trained in the scientific method, Freud had, as he has told us, only one God: reason. It was for him a God whose ways lack the flamboyance and the omnipotence of the gods of religion. "The voice of intellect," he wrote in *The Future of an Illusion,* "is a soft one, but it does not rest till it has gained a hearing. Finally, after a countless succession of rebuffs, it succeeds. This is one of the few points on which one may be optimistic about the future of mankind."[27] It is "the primacy of the intellect," then, which man can look forward to in the future, for in intellect alone, making use of science, can any hope be held out.[28] But when? Not now, Freud says, for the movement of reason is slow; reason is a god who provides no instant gratifications, and who, when he grants us our wishes, does so within the limits of nature alone. Following him, we can hope for progress, not necessarily for ourselves, but for those who come after us. If the prototypical style of the preacher and the prophet is to set man's heart aflame, by angering, arousing and inspiring him, Freud's style is just the opposite. Even when he praises his own God, reason, urging that all men pay homage, exclusive homage, to him, he makes no attempt to exaggerate the gifts this God will return to his worshipers. Freud lived by the scientific method, the finest expression of reason, but listen to the modest way he commends even his own most sustaining faith: "We believe that it is possible for scientific work to gain some knowledge about the reality of the world, by means of which we can increase our power and in accordance with which we can arrange our life."[29]

Two questions are raised by that quotation. The first is the extent to which it is even rational to think that the scientific method is the only way of gaining some knowledge of reality. There is no need here to rehearse the long argument between the proponents of the narrowest possible scientific rationalism and those who hold it possible that knowledge can stem from many sources, not all of them "rational." Clearly enough, people believe they learn something from art, music, philosophical speculation and experience; modern science has grown in its power and sway, but it has not succeeded in driving out all other claimants to knowledge. On the contrary, it is precisely the inadequacies of a purely scientific understanding of the nature of things which have kept other modes of knowing alive and active. Moreover, science has not provided any clue whatever about the way in which scientific knowledge can or should be used in constructing a culture, developing a full system of values or enabling us, in Freud's own language, to "arrange our life." This is not to say that attempts are not made periodically to move from the "is" of scientific rationalism to the "ought" of a life of culture and morals, but the history of such attempts is replete with failure. Thus the second question which must be put to the Freudian rationalism: What are we to do with the scientific knowledge once we have it? Freud gives us no real answer to that question, a question which clearly implies that somehow scientific knowledge to be of any final value must be integrated into some kind of value system. Freud leaves us high and dry at that point.

It is the presumption of the righteous that the prophet is the most feared man in a contorted society. He calls that society to accounts, sets before it the path of righteousness, and reminds it of the myths of happiness which are part of its primeval possession. He is not thanked for this service; he is likely to be killed for rendering it. But he is the hero of the righteous, who need him and believe that society needs him. Yet as intolerable as the prophet may be, there is no room whatever for the antiprophet. Even the righteous recoil from him, and especially so, for they live in a world which has meaning and makes sense. They too say the way is hard, but they think that there awaits the seeker what eye has not seen nor ear heard; so it is worth all the trouble. But the anti-prophet says both that the way is hard and that, in the end, there is even then not much to hope for. Justly is he feared, for if the old question, "What can I hope for?" is answered the way Freud answers it—well, don't hope for much and not now—then the dissolution of all myths seems at hand; and when they go, hope goes. "Freud the fink" has become one cry of the day—that Freud who deflated the pretensions of just about everyone, especially those dedicated to a political solution to man's woes. Even those who have followed Freud, willing enough to make him their master, have tended to draw back from his analysis of civilization. Who can stand the gloom? As Paul Roazen has written, "The revisionist tendency is an intellectually amiable one. The post-Freudian literature has none of the disturbing quality of Freud's own writings. Revisionism has a soothingly reassuring tone. Instead of Freud's complex grasp of the way culture both thwarts and fulfills human needs, the revisionists substitute a less tragic view of the relation of man to society. Change modern civilization, they advise, and man's naturally harmonious nature will be able to unfold itself. The notion that neurosis is due to mental conflict tends to disappear; it is the environment which needs changing."[30]

The question which needs an answer, and no more so than at present, is whether there is anything to hope for and some way of salvation worth pursuing. We may dismiss the objection that there *must* be something to hope for, else why live, why not here and now do away with ourselves? Philosophers, on occasion, speak as if the alternatives must be put in this way, and so sometimes do theologians. But it does not seem as if most people do. They either find something to hope for, or they make the most of hopelessness, hanging on until death. And one can of course spend one's life searching for reasons to hope; even if they are never found, time is passed, and that too is something. None of these strategems, however, does away with the question: What is there to hope for? Every generation has to face it anew. Where should one begin? History is studded with ways of salvation, most of them hopeful; by definition, they would strike no one as salvation did they not proffer hope. To choose Freud's analysis of civilization as a point of departure does not appear to offer many benefits. There are happier figures to be found, and more congenial theories—congenial, at any rate, to our wishes. The advantage of beginning with Freud is that we start at the bottom of a pit, caught in an elaborate snare. If we can break free from the snare, and if we can climb the slippery walls of his pit, then we will have accomplished something. There cannot be many harder intellectual tasks left after that. And if we cannot escape, what then? What if the "reality principle" proves too strong for us? Something may turn up.

A formal way of putting the problem is to ask whether reality is amenable to the aspiration for transcendence. Two kinds of transcendence, interrelated, come into play. Can we transcend our human natures, which put so many obstacles in our way? And can civilization transcend those human natures which cause it so much turmoil?—and that means all human natures. The word "transcendence" can be understood in a variety of senses, but the sense I will make use of is that of "going beyond"—going beyond human nature as we now know it, and going beyond civilization as it now presents itself to us. One can "go beyond" in a weak and in a strong sense. In the weak sense, a higher, more satisfying plateau is reached, but that plateau remains part of the known or readily conceivable landscape, examples of which are now and then to be found even on the surrounding plains. In the strong sense, one leaps beyond everything now known, going beyond familiar fantasies; the difference becomes radically qualitative and not just progressively quantitative. Only the weak sense, following Freud, will concern me; it is the only one I believe can be understood.

The relationship between the quest for a tolerable human nature and the search for a viable civilization are joined, as Freud's writings would imply, by the simple fact that the latter can be made up of none other than tolerable people. A civilization, whatever else it may be, is a human collectivity. Do away with people and there is nothing left, no problem at all. Since in any imaginable culture people will remain dependent upon each other in one way or another, and there thus will remain present the seedbed of mutual aggression, a transcendence of both present human nature and present civilization will remain a single task, though with perhaps different aspects depending upon whether we are dealing with individuals or groups.

The great value of Freud's reflections upon civilization is that now, more than ever, they provide one with an ultimately harsh and unsentimental way of looking upon technological society and the human substrate which underlies it. That kind of a starting point is imperative, if only in order that the development of a science of technological limits can take its point of departure from pessimism rather than optimism. For the problem in the past has not been to see the good side of technology, but rather to have a way of confronting its harmful side. But that requires a theory of culture, one which will provide some insight into the painful fact that technology goes awry. It is impossible to cope with the fact apart from a theory of human nature which refuses to see the aberrations and mistakes of technological societies as only accidents and coincidences. If the result of this refusal is some degree of pessimism, then this may bespeak more the way we have allowed our hopes to run—for pessimism can only spring from misplaced hopes—than the actual nature of genuine possibility, not all of it doleful.

The advantage of using Freud's theory of civilization as the base point of a social ethic may be less evident. On the face of it, Freud presents few ingredients to construct an ethic, either of life in general or technological life in particular. As suggested earlier, it is difficult to see how a move is to be made from his harsh scientific rationalism to an integrated view of the world which allows human aspirations to be recognized, value systems created and ethical codes devised. It will not do to see those aspirations as only sublimations (which trivializes them), all value systems as only adaptative modes of existence and all ethical codes as sheer repressive devices to keep the instincts of individuals within socially tolerable boundaries.

Yet I want to contend that the problem is less one of seeing how one can generate a social ethic from Freud's theory of civilization and its discontents, than of seeing how advantage can be taken of that theory in trying to make continued use of an ethic which has traditionally rested on a more benign and optimistic view of human life. There are a variety of ways in which that can be done: by persuading us, first, that any social ethic must build within it a healthy sense of the perennial struggle between individual and society (even if the struggle is cast in terms more hopeful than Freud allowed); second, that a good social ethic is one which restrains its hopes for a human nature that will suddenly be transformed and works, rather, with a sense of permanent tension and conflict; and third, that in approaching technology it will look not only to the possible good of technology but will have a sharp eye out for those ways in which the lure of technology falsely promises to liberate human beings from nature, the body and other humans. Freud is the sharpest critic in modern history of false liberations. An abiding sense of the difficulties of any and all kinds of liberation, combined with a tentative yet firm willingness to see that life need not always remain the way it first presents itself, provides the best possible ground for taking an inherited ethic into the troubled waters of technological change.

## NOTES

1. Sigmund Freud, *The Future of an Illusion,* trans. W. D. Robson-Scott (Garden City, N.Y.: Doubleday, 1961), p. 73.
2. *Ibid.,* p. 81.
3. Sigmund Freud, *Civilization and Its Discontents,* trans. Joan Riviere (Garden City, N.Y.: Doubleday, 1958), p. 105.
4. *Ibid.,* p. 17.
5. *Ibid.,* p. 16.
6. *Ibid.,* p. 25.
7. *Ibid.,* p. 29.
8. *Ibid.,* p. 41.
9. *Ibid.,* p. 43.
10. *Ibid.,* p. 61.
11. *Ibid.,* p. 63.
12. *Ibid.,* p. 65.
13. *Ibid.,* p. 66.
14. Sigmund Freud, *Beyond the Pleasure Principle,* trans. James Strachey (New York: Bantam Matrix Edition, 1967), p. 67.
15. *Ibid.,* pp. 70–71.
16. *Ibid.,* p. 74.
17. *Civilization and Its Discontents, op. cit.,* p. 70.
18. *Ibid.,* p. 71.
19. *Ibid.,* p. 74.
20. *Ibid.,* p. 75.
21. *Ibid.,* p. 82.
22. *Ibid.,* p. 89.
23. *Ibid.,* p. 99.
24. *Ibid.,* p. 100.
25. *Ibid.,* p. 105.
26. *Ibid.*

27. *The Future of an Illusion, op. cit.,* p. 87.
28. *Ibid.,* p. 88.
29. *Ibid.,* p. 90.
30. Paul Roazen, *Freud: Political and Social Thought* (New York: Knopf, 1968), p. 268.

# CHAPTER 3

# *Modes and Manifestations of Technology*

FOR FREUD, the problems of technology are clearly subordinate to the more fundamental issue of the social neuroses endemic to civilized life. Though he recognizes the "dejection" brought on by man's power to destroy himself, and the failure of human beings to find happiness despite their self-made likeness to God, at no point is he prone to blame technology for the present human condition. At most, technology may exacerbate the feeling of helplessness and unrest, but it is by no means the crux of the human dilemma. On the contrary, Freud is at times prone to be lyrical on the benefits of technology, which makes important contributions to the need for "beauty, cleanliness and order."[1] Moreover, there is no indication that he drew what is by now the common distinction between science and technology. Science remained the handmaiden of his God, reason, and its benefits are as much practical as theoretical, in hygiene, agriculture, irrigation, communications, metallurgy and animal husbandry.

The primary hazard Freud perceived in technology appears limited to one area only, the destructiveness of modern weapons of warfare, which enable human beings to express their aggressiveness on a far wider and more devastating scale than was ever possible to earlier generations. In the triad of the sources of human suffering—the body, nature, and other human beings—it is the last which is made worse by technology; Freud did not see technology as contributing anything other than benefit in the struggle against the ills of the body and the assaults of nature.

This optimism does not contradict Freud's more general pessimism about the possibility of happiness; he never claims that man will be *saved* by technology. But it does underline important differences between Freud's pessimism and that of contemporary counterculture critics. Where the latter (cf. Theodore Roszak) are apt to lay the blame for human misery squarely at the feet of technology and the rationalistic temperament which underlies it, Freud never singles out technology for any special blame nor indicts rationalism as the chief villain of modern life. The genius of Freud lay in his drive to penetrate to the deepest sources of civilized unhappiness, which he perceived antedates the advent of modern technology and which (as in his critique of Communism) is inherently skeptical of social and political cure-alls for a sickness which is an inherent part of human nature. It is hardly any wonder, for this reason, that Freud's radically antiprophetic psychological analysis would be passed over (other than in its distorted, attenuated and revisionist forms) by modern gurus, who must if they are to retain the title (that is, the celebrity) produce the promise of happiness and salvation.

Nonetheless, though Freud, by taking the sweep of human history for his analysis and not just modern life, was able to see deeply and soberly the magnitude of the human problem, it is possible that he failed to perceive that technology can introduce far more hazards into life than simply modern warfare. While one suspects he might quickly have granted that one of the allures of technology for the individual is its seductive song of total gratification of all desires, it is not clear he would so readily have conceded the power of technology to impede the power of altruism, which drives toward a transcendence of individual egoism into a merging of human beings with each other. The most powerful and persuasive contemporary critique of technological society has centered on its power to alienate and isolate the individual. While this critique must in part presuppose that somehow individuals in pretechnological societies were less alienated, more at one with each other—a point which Freud's analysis of all civilized life denies—it has more pointedly centered on the special capacities of technological societies to create anomie and alienation. The modern production line, the mass and complexity of technological artifacts, the distancing of individuals from the workings and control of the overall technological system, the interchangeability of human parts, the professionalization of crafts and disciplines, the anonymity of a technologized urban life, together conspire to separate human beings, to make altruism seem a pointless gesture, community an ever more distant dream.

Moreover, as I will develop more fully in later chapters, technological societies impose both a *tyranny of survival* and a *tyranny of individualism.* They impose the former because, in times of stress, their extreme fragility (stemming from the high base of expectation they engender and the high degree of total control their complexity demands) is instantly and terrorizingly apparent, creating a natural environment for an obsessive fear of annihilation, i.e., a tyranny of survival. They impose the latter—monomaniacal individualism—because only the privatized life seems viable or endurable in the midst of a system which presents itself as impersonal and uncontrollable. Thus is intensified the tyranny of individualism, which demands that each person create his or her own world *ex nihilo:* self-direction, self-realization, self-fulfillment—self, self, self.

If Freud underestimated the destructive powers of high technology—which in any event only made its appearance as a mass phenomenon after World War II—contemporary analysts (particularly the most hostile critics) have failed to do justice to it because of a penchant for abstraction and generalization, trying to grasp as a whole a subject which is not a whole but rather a congeries of pieces and parts by no means integrated into a system. The result of this misplaced holism has been either utopian epiphanies of technological fervor or vitriolic condemnations; each has provided some illumination, but not enough. The requirement at this point is for an understanding of technology which will neither idolize nor condemn it, but instead will be able to grasp its positive values, perceive its present destructiveness, and estimate its future possibilities and hazards. The point of this exercise is not more theorizing about technology—which makes little difference one way or the other in its actual impetus and manifestations—but as a way of establishing a foundation for the "science of limits" which, as I argued in the opening chapter, is at present needed.

The first step in such a science is to face the most evident "psychological reality principle" of technology. That principle can be stated in a direct and even dogmatic form. Whatever the ragings of the present antitechnological counterculture, whatever the hesitations and anxieties of their more measured and established counterparts in "straight" culture, whatever the chilling effect of daily and quite accurate messages on the latest depredations done to man or nature by technology, whatever the revolt at an impersonal, mechanical socio-industrial order—no matter what, technology as a human phenomenon, and roughly in its present state, is here to stay.

At least for the "developed" nations of the world, life may become better or it may become worse, but either way it will be a better or worse technological life. And there is every prospect that the developing nations of the world will become more and more technological as well.

To work with any other hypothesis is to indulge in day-dreaming. For rhetorical purposes, and to catch the ear of a society inundated with uplifting, neutralizing or depressing messages (a leading feature of modern media), wholesale condemnations of technology may have their place. But for purposes of coping with reality, they need to be seen for what they are, pious hopes, mainly appealing to a very small number of the affluent. I posit no inevitability of history here, as if there could be no other possibility. I posit only history, together with the massive reality before our eyes, a reality which shows the very deep and tenacious grasp of technology, even among those who would do away with it (who, it has been noticed before, are rarely apt to forgo the efficiency of printing presses, tape cassettes, television and radio in propagating their antitechnological message). If it is not quite possible to say that like God technology always *was,* it is possible to say that *it is* and always *will be.*

The roots of the psychological reality principle of technology are *not* to be sought in present manifestations of technology. On the contrary, though it is possible to note historical milestones in the historical development of science and technology, one thing seems evident in that history taken as a whole: man has ever been a technological animal, a builder, a shaper and a tool-maker. The fashioning of tools and implements, the essence of technology, has provided, from a zoological and anthropological perspective, an essential criterion for distinguishing humans from animals. It is also, from a philosophical perspective, a key sign that one is dealing with intelligence and rationality. If one wants to look for an ultimate source of the technological fallacy—that man must by nature become technological—then one could find no better place to start than with the assumption that the making of tools and implements is a *sui generis* to man, part of what it means to be *Homo sapiens*. Yet there is every reason to think this is no less than the literal truth of the matter. This truth is by no means hostile to the parallel truths that to be human means to think abstract thoughts, write poetry, experience emotions, paint pictures, compose music and build cultures. But to emphasize all of the latter while ignoring the former is simply not to have in hand the full truth, part of which is that *man is by nature technological.*

How could it be otherwise? Man has a body which must find ways of surviving in a variety of environments, a brain which can think and set goals, most of which require a material manifestation, an imagination which can hope for some relief from the ills of the body, a will which can seek, and passions which can desire to build, create and improve. And man lives amidst nature, the only stuff besides himself that he can and must work and live with. Man is fashioned by nature and he fashions it. There is no other choice.

The problem is how we are to live with nature, that nature which is the external world and that nature which is human nature. Technology is not just *a* component of that problem, one among many; it is the essential component. The question now is, how much of it do we need? How can we control our dependency upon it? How can we make it serve rather than enslave us? None of these questions would arise were we not *and of necessity* wedded to it, not because of the kind of culture we live in, but more fundamentally simply by virtue of being human beings. Our culture exacerbates the technological problem, but it is our human nature, with its needs and demands, which had no choice but to create it. The very meaning of "civilized life," which Freud did *not* underscore, is that it is a technological life, however primitive the technology.

I stress these points to do away with a false, but very popular dualism, one which abstracts man, on the one hand, and technology, on the other, as if the two were quite separate kinds of realities. It is man's nature to be technological. He cannot be otherwise without ceasing to be human. The dualism is dead; stillborn. When we speak of technology, this is another way of speaking about man himself. The question is not one of Yes or No, affirmation or denial of technology, but the harder question of *how much* and *when* and *in what circumstances*. Since we are by nature technological, we must learn to live with ourselves as inherently technological, not only that we might better learn how to enjoy and profit from technology, but also that we learn how, in its more complex, confusing, destructive maturity, to keep ourselves from being killed by technology. In other words, to learn how to be healthy rather than sick technological selves.

Yet if it is the disasters of technology which have opened eyes to a fuller (and more doleful) meaning of the potentialities of our technological selfhood, now historically older and ripened, this does not nullify the observation that man is and will be inveterately technological. But how can he be technological and wise at the same time? At the very least, the disasters of technology show why it is a mistake to set false gods before us. The false god of a consciously "modern" technology, a child of the enlightenment and Baconian muscularity, promised human perfection and salvation. Man was to find, with science, full knowledge. Through technology, he was to put that knowledge to work in ridding the human condition of the finitude of the body and of nature. That god is dead, even if the corpse is still twitching. We are now in the more pedestrian position of seeing that, while sacred cows make poor gods, they may be edible, even tasty, if care—extreme care—is taken in the cooking.

I have claimed that the problem of science and technology is one of degree, putting aside the views of those increasing number of jeremiads who see technology as the doom of us all. If they are right, it is too late anyway; if they are wrong, they are just misleading everyone. To confront science and technology as a matter of degree—by which I mean in terms of comparative goods and evils, and as a function of motive, result, place, time and circumstance—it is necessary first to take a look at some of the human needs and motives which lie behind its creation, continuation and expansion.

It ought to be evident, in the first place, that human beings have always desired to know what they and the world are all about. Science is no more than an expression of this desire, providing a particularly effective set of methods for coming to an understanding of things as they present themselves in their empirical givenness. The desire to know is another meaning of what it is to be human. Moreover, that desire has the additional impetus of a *need* to know, for the sake of survival, adaptation and control. It would take a lot of nuclear weapons to do the damage dear old Mother Nature has done by barely lifting a finger. That the human population did not reach one billion until 1850 tells us something about the kindliness of her methods of birth control. While the underdeveloped countries have increasingly nasty feelings about the rich nations, there are very few inhabitants of those countries who do not prefer, given the chance, to exchange their natural calamities for our manmade kind. It is still better to suffer from alienation in the United States than be wiped out by a hurricane in Pakistan. The development of science is expressive of the long-standing human need to master nature, to know and control it. We may speak as much as we like of the need to live in harmony with nature, of the need to develop an attitude toward nature which gives it its due respect as the source of our nurture. But we should still be sensible enough to realize that it is easier to live in harmony with a tamed than an untamed nature, and that science and technology provide the best means we have to do the taming.

It ought to be evident, in the second place, that disparate kinds of technologies have sprung from variable human needs and desires. Most discussions of technology are far more confusing than they need be because of a failure to distinguish among variant kinds of technologies and the diverse kinds of needs they have historically been developed to serve. Technology is not just one lump phenomenon; it is a whole collection of items and events. Nor is the meaning and impact of technology—its human significance—the same in all times or in all places. The advent of nuclear weapons was a technological disaster, and so was the advent of excessive population growth and the introduction of DDT into the environment. But they are different kinds of disasters, bespeaking different causes, motives, results and histories. To lump them all together as uniform examples of the ills of technology explains little and clarifies nothing.

I want to attempt a method of classifying technologies according to the different kinds of needs and desires they were or are meant to serve. This may help us to better analyze and evaluate the nature of those needs and desires, by looking at some of the technological ways they are expressed; and it may help us see more clearly just what is at stake when we try to measure the comparative gains and losses brought about by particular forms of technology. Five broad categories of technology can be distinguished: (1) *preservation technology*—meant to serve the human need to survive in some meaningful way; (2) *improvement technology*—designed to allow man to improve his condition according to his imaginative, spiritual and sensual visions; (3) *implementation technology*—designed to make systems, whether social, scientific or economic, function better or more efficiently; (4) *destruction technology*—designed to totally control or utterly get rid of that which man does not want; (5) *compensatory technology*—designed to ameliorate the physical or psychological harms done by technology.

Within each of these categories, further divisions and distinctions are possible (see Table 1).

## Preservation Technology

Preservation technologies are developed to better enable man to meet his need to survive the rigors of nature, enabling him to tame, to adapt to, and to profit from it. The most fundamental of these may be referred to as *adaptative*. They are technologies whose purpose is adaptation to environment, a basic requirement of physical human existence. Primitive farming tools, at one extreme, and space suits, at another, are typical of such technologies. Irrigation systems, fertilizers and air conditioning are also examples of adaptative technologies. The history of these technologies is that of an ever-widening range of adaptations and the pacification of all environments, including the most hostile, for human use and habitation. Once it was established that, with enough ingenuity, any environment could support human life, an extension into those environments became inevitable; an increasing population has made the extension necessary.

Closely related to adaptative technologies are those which are "protective," whose purpose is to ward off or neutralize environmental and natural hazards. The presence of dangerous animals meant that primitive man had to develop weapons of protection. Extremes of cold and heat meant that housing had to be devised. The same principle has been at work in the development of medical technologies. Man is subject to a variety of hostile germs and diseases, various routine but nonetheless lethal afflictions of the flesh. Survival has required that they be attacked and neutralized.

"Corrective" technologies, by contrast, have developed out of the intuitive or implicit assumption that there is a standard of human normalcy—a set of bodily, mental and emotional characteristics which human beings need in order to survive at all and to be positioned to develop their full natural potential. Thus eyeglasses presume that people need to see, and wooden legs that they need to walk. The artificial introduction of vitamins and a wide range of corrective medical technologies focused on disease and dysfunction, are aimed at the establishment of a certain minimal level of bodily functioning. Even though negative eugenics is still in its infancy, it would fall within this form of technology, aiming to rid human beings of those genetic characteristics which make them less able to survive or adapt. Tranquilizers and various forms of chemotherapy do for the mind and emotions what various medicines and prosthetic devices do for the body, enabling them to function in a "natural" or "normal" way. The philosophically important point about corrective technologies—often employing the "medical model" for their criteria—is their presumption of a standard of normalcy, epitomized in the notion of a "healthy" person, sound of mind and body. Since nature rarely exhibits such a person, and then only fleetingly, the notion itself must be judged a human construct, which has undergone continual historical modification, with an ever-higher, escalating standard of the "normal" and the "healthy."

"Investigative" technologies form a special class within preservation technology. They are technologies in the service of science, whether that science be directed simply toward knowledge per se, or knowledge in the service of various

TABLE 1

## Varieties of Technology

### PRESERVATION TECHNOLOGY

| *Adaptative technologies* | *Protective technologies* | *Corrective technologies* | *Investigative technologies* |
|---|---|---|---|
| farming tools | spears, axes | eyeglasses | telescopes |
| hunting implements | antibiotics | hearing aids | microscopes |
| wheels | vaccines | wooden legs | linear accelerators |
| irrigation systems | housing technologies | prosthetics | amniocentesis |
| dams | paints | vitamins | x-rays |
| weather controls | pesticides | negative eugenics | EEGs and EKGs |
| fertilizers | herbicides | insulin | |
| contraceptives | | tranquilizers | |
| space suits, diving gear | | euthenics | |
| automobiles | | | |

### IMPROVEMENT TECHNOLOGY

| *Enhancing technologies* | *Decorative technologies* | *Aesthetic technologies* |
|---|---|---|
| genetic engineering | nonbiodegradable bottles | redwood housing goods |
| psychedelic drugs | plastic surgery | stereo sets |
| automobiles (high-powered) | auto decor | records |
| electric stimulation of brain | cosmetics | |
| space ships, SSTs | paints | |

## IMPLEMENTATION TECHNOLOGY

| *Facilitating technologies* | *Economy-enhancing technologies* |
| --- | --- |
| abacuses | planned obsolescence |
| computers & data banks | advertising technologies |
| information systems (radios, TVs) | consumer-oriented technologies |
| cans, boxes, containers | research and development of new products |
| transportation system (trucks, planes, trains) | automobiles |
| typewriters | space technology |
| tapes | |
| telephones | |
| tool-making machines | |

## DESTRUCTION TECHNOLOGY

| *Manipulation and Control technologies* | *Obliteration technologies* |
| --- | --- |
| wire-tapping and snooping devices | nuclear weapons |
| brainwashing technologies | defoliants |
| | biological and chemical warfare |
| | vacuum aspirators |

## COMPENSATORY TECHNOLOGY

| *Psychological Compensation* | *Physical Compensation* |
| --- | --- |
| tranquilizers | reprocessing technologies |
| Muzak | sewage treatment plants |
| technologically assisted vacations | chemical fertilizers |
| encounter and sensitivity techniques | "garden" apartments |
| hiking, skiing and sports equipment | |

adaptative, protective or corrective goals. A telescope would be an instance of the former, while a microscope would be an example of both the former and the latter. Historically, the earliest scientific methods were of a gross trial-and-error kind, followed by more sophisticated concepts of the scientific method. The great leaps of science have come with the invention of exquisitely conceived and engineered devices for discovery and experimentation, together with mathematical models. Modern science has heavily depended upon technological devices for its advances. It is a mistake to draw too sharp a distinction between science and technology. While it is true that the application of scientific knowledge is what we normally think of as technology, the fact that the accumulation of scientific knowledge is so heavily dependent upon investigative technologies casts the distinction in a different light. Science makes technology go forward, but without technology science could itself not progress.

## Improvement Technology

I mean by "improvement technology" those technologies designed to enable man to meet his felt need to project higher possibilities, to go beyond what would be the ordinarily perceived limits of human nature as given, even that nature which has had obvious defects removed. The most important forms of such technology I will call "enhancing" technologies. Genetic engineering, especially positive eugenics, is a case in point. In its more visionary aspects, it projects an improved or "better" human nature, a "higher" form of man. Psychedelic drugs are often defended out of similarly utopian motives—a higher human consciousness can be created, opening up radically new forms of experience. Electrical stimulation of the brain is embraced by some as preparing the way for the rational control of human behavior and the eventual eradication of crime and war, both historically endemic to human nature as given. At a more mundane level, high-powered automobiles are developed to offer the possibility of bringing otherwise unattainable fantasy satisfaction (sexual or mechanical).

The philosophical assumption behind "enhancing" technologies is that man can become something other than he has been historically; he can control either his own evolution or consciousness in a way that will enable him to go beyond that which is presently accounted "normal" or possible. The fine line between "corrective" and "enhancing" technologies is obvious. "Corrective" technologies, as noted, have shown a history of escalation, the establishment of ever-higher norms of "healthy" or "natural." Since nature only rarely produces that which would meet a perfect prototype of "normalcy" or "healthy," man has stepped in to invent and re-invent the norms and models.

If it is granted culturally that nature can and may be corrected (in the sense that everything which happens in nature is not necessarily good for man), the ground is laid for improving upon nature; the acceptance of the former is usually taken to legitimate the latter. Indeed, a *correction of* and an *improvement upon* nature become very difficult to separate, particularly when undergirded by the belief that it is man and not nature who should establish the appropriate norms and directions.

"Decorative" technologies are expressive of a related phenomenon. If it is possible to envision radically "higher" modes and states of human life, it is even easier to imagine and implement more humble improvements and alterations. Cosmetics can be used to enhance the possibility of improvements in human beauty. A variety of automobile decors can be developed to suit different tastes. Plastic surgery can be used to improve, even better than cosmetics, the faces nature provides us with. In this general respect, it is possible to distinguish between "decorative" and "aesthetic" technologies. The "decorative" trade upon highly variable, subjective tastes and fashions; it would usually be admitted that little more is at stake. In the case of "aesthetic" technologies, however, it would probably be granted that beauty is an intrinsic value, and that technologies which make beauty possible or accessible have a high claim to efforts toward development. Thus stereo sets and records, which bring us operas and concerts, or the use of modern chemicals to preserve old paintings, might be seen as facilitating a transcendence of mere taste and fad.

## Implementation Technology

I mean by "implementation" technology those technologies designed to facilitate a more efficient, productive, successful working of social, scientific or economic systems. "Facilitating" technologies are second-order instrumentalities, assisting first-order systems. Abacuses once helped people to count more rapidly and efficiently (and in some places still do), while computers are wonderfully efficient for correlating, computating and storing information. Those information systems represented by radio and television bring news, close distance gaps and bridge cultures. Cans, boxes and containers enable us to store things, while transportation systems enable us to move both people and goods. Although "facilitating" technologies may become ends in themselves, their usual function is that of instrumentally enabling us to achieve other ends. An oxcart can be used to move cargo, but airplanes are faster and carry larger quantities. I can walk to my neighbor's house to tell him something; it is faster to call him on the telephone.

"Economy-enhancing" technologies are those designed to

give people work to do, to raise gross national products, and to achieve certain standards of living. Planned obsolescence serves all three purposes, as will the development of new products, the constant changing and "improvement" of old products. At least they will serve these purposes in certain kinds of economies, especially those of a capitalist variety. The great attraction of such technologies is that they trade upon some basic human needs and desires for self-preservation and self-enhancement.

## Destruction Technology

"Destruction" technologies serve the need of man to either totally control or totally destroy that which he perceives as a threat. "Manipulation and control" technologies place the emphasis upon mastery of other human beings, so complete that they can pose no hazard to the achievement of one's own or societal goals and values. Wire-tapping and snooping devices are illustrative of technologies which, by invading privacy, provide us with knowledge of otherwise hidden motives, plans and acts of those we consider dangerous. Once they are known, steps can be taken to control them. Brainwashing technologies go even further. They get inside of another, enabling us to do with him what we will.

"Obliteration" technologies have a special place. They are designed with one end in mind, to utterly obliterate or otherwise remove threatening forms of life. Nuclear weapons have as their sole purpose the killing of human beings, in as massive and comprehensive a way as possible. Biological and chemical warfare, in its most lethal forms, has the same end as nuclear weapons. Defoliants are designed to wipe out various forms of vegetation. Vacuum aspirators remove unwanted fetuses. What distinguishes "obliteration" technologies from all others is the very narrow range of purposes they can fulfill and the very wide, lethal effects of their results.

## Compensatory Technology

These are technologies designed to provide physical or psychological compensation for, or amelioration of, either the harms done by technology or the special pressures it generates. Typical compensatory agents would be tranquilizers (to reduce the tensions induced by a fast-paced technological society), Muzak (to ameliorate the tedium or confusion of technological environments), encounter and sensitivity techniques (to reduce alienation and anomie), and elaborate hiking, skiing and sports equipment (facilitating periodic "relaxation" after the stresses of technological living). Among the physical compensatory technologies would be reprocessing technologies (to cope with trash and conserve resources), elaborate sewage treatment plants and systems (to ward off pollution and to deal with the inadequacies of natural systems to absorb technological waste products), and chemical fertilizers (to replenish depleted natural nutrients in soil).

The need for compensatory technologies might not exist at all were it not for the scale of magnitude of all other existing technologies, which create a host of hitherto unknown physical and psychological pressures. By definition, a technological society will try to find technological solutions to the problems created by technology. While it might be argued that such a process is self-defeating (more of the same, when the problem is already more of the same), that case becomes more difficult to make when, as in the case of sewage or trash, there is simply no longer any possibility of natural disposal in the very near future without incurring immediate disaster. And while it might be contended that the heavy societal dependence upon tranquilizers is a bad, because symptomatic, method of handling large-scale depression and anxiety, it may in fact be impossible to use more natural (e.g., psychoanalytic) techniques because of the very quantity of those who need treatment. Though it may well be both possible and reasonable to conceive of the further hazards introduced in the long run by a dependence upon compensatory technologies, the pains and problems of the immediate present militate for quick, efficient solutions, simply in order that day-to-day life may go on. That there is a vicious cycle here is of course obvious.

I do not put these categories and lists forward as either definitive or wholly coherent. Endless games can be played trying to decide which technological item belongs under which heading. The problems in creating the right kinds of categories and lists are, in fact, suggestive of why it is difficult to evaluate technology. For the wholehearted opponent of nuclear weapons, they are properly classified as a destructive technology. For those who see them as a protection of the "American way of life," they will be classified as a protective, defensive technology. Many who would oppose nuclear weapons because of their exclusively destructive characteristics heartily welcome vacuum aspirators, which enable physicians to perform quick, safe, cheap abortions, designed to protect us against unwanted children. And just what is an automobile designed to accomplish? For those who live in areas where people are spread out, they are a way of adapting to environment. For those who live in highly industrialized societies, they are an economy-enhancing technology. For those seeking privacy, they are an alternative to use of public mass-transportation systems.

The same technology can serve a variety of very different needs and desires, and thus be interpreted very differently depending upon what goals one sees them as achieving. That reality raises one of the fundamental problems in trying to understand technology: its multi-effects, and multi-significances. Few if any technologies have one human meaning only, any more than they have only one cultural significance.

Moreover, technology has its own logic and this logic is one with human logic, where step *a* implies step *b*, and *b* implies *c*. . . . It is perfectly sensible for starving people to create tools with which to till the land; they are adapting in order to survive. It may seem no less sensible, once the logic of survival has been introduced, to extend the means of adaptation as far as possible, to the habitation of planets and the depths of the sea. Moreover, the next step is all the more likely to be taken when, as in the case of moon rockets, it is economy-enhancing as well. And why does it make any less sense, if we have enemies who would kill us, to try to kill them first and more efficiently; nuclear weapons work better against large populations than guns. If it is reasonable for an Indian culture to manufacture some pottery to sell in order to eat, why is it any less reasonable for an industrial culture to make automobiles to sell in order that it might eat?

Of course we think we know why some of the new technologies are very different from the old; though the logic is sane at one stage or place in human history, it can become insane at another. We do not happily lump pottery-making and auto-making together. But it is difficult to see that anything totally different is going on so far as the satisfaction of needs is concerned. Though the needs being served are more sophisticated and differently conditioned in auto-making than in pottery-making, the motives behind both production technologies may have much in common. That auto-making is hazardous to the environment while pottery-making is not may be of secondary importance at the level of root motivation. People are born into cultures which, for them, are a given preexisting environmental reality. To eat, they must do that which the culture requires them to do as the price of obtaining food. In our culture that means manufacturing automobiles; only a few can eat by making pottery. What is important, in the first place, is that environmental resources must be used for adaptation to the culture in which one finds oneself. That they may be ill-used is an inevitable possibility if used at all.

The importance of distinguishing, if possible, between true and false needs becomes critical at this point, particularly to make sense of the fact that technology seems to engender "needs" which might not otherwise exist and which appear not to have existed in earlier eras. There now exists no technology which does not serve some expressed human need. The earliest technologies, which fell heavily in the preservation category, were only the precursors of the later; and it would undoubtedly be easy to trace the later, more advanced technologies back through their genealogical trees to earlier, less advanced technologies. More difficult to understand is why the process seems to go on without end, why every step either leads to or suggests the next step, and why those steps are taken. Why, to bring the question to a point, are human beings never satisfied and why, more than that, do they go on with their technology even when they become well aware, first, of the diminishing return from more technology once basic needs have been satisfied and, second, of the rapidly expanding range and depth of the hazards which the "more" carries with it?

The simplest and most plausible explanation is that human beings have more needs and desires than any technology has yet met, and that the satisfaction of one need, far from completing man, makes possible attempts to satisfy still other needs. Human beings have desires which are infinite, knowing no bounds of hope or aspiration. In practice, short of sheerly physiological criteria, there is no effective way of distinguishing between "needs" and "desires." All desires may not represent needs, but all needs carry with them desires; and a strong enough desire will have all the psychological characteristics of needs. Who is ever completely satisfied in all things, and for how long? Technology is a gamble that infinite desires can be satisfied by finite means.

One way of defining a "false need" is to stress its artificiality, trying to show that it does not arise in man's "natural state," that it must be induced from outside—for instance, by life in a technological society. But it is notoriously difficult to find man's "natural state," and even if emphasis is laid, say, on the way media manipulators can induce fears, anxieties and desires designed to lead people to want or buy things which apparently only technology can provide, that still does not explain why they are so readily susceptible to such manipulations, a susceptibility which must have some foundation in real needs. For all these reasons, I doubt that one can profitably speak of a "false need" as such. The emphasis is better placed on false satisfaction of real needs. The satisfaction could be false either in the sense that technological societies exacerbate ordinary needs, and then offer more technology as a solution to the new problems thus created, or in the sense that false or harmful objects of satisfaction are offered. A jet trip to Bermuda may be enjoyable, but if it is offered as the solution to the intensities of technological life, the possibilities are more illusory than real.

The key to making the needed distinctions is to determine when real needs are being falsely satisfied. How is that to be done? Let me offer a formal definition. A real need will be falsely satisfied when (*a*) the continuing existence of the need, not actually satisfied at all, is obscured or hidden by the immediate pleasures, or surcease from pain, produced by a technological artifice; and (*b*) when, despite any confirmatory evidence, technology is understood to hold out the hope that, given enough time, it will be able to satisfy all needs. The point of the first part of the definition (*a*) is that it is possible to simulate the satisfaction of needs, and effectively enough so as to deceive people into believing that the needs have been satisfied. The point of the second part of the definition (*b*) is that unsatisfied needs can be more easiiy tolerated if there is hope that, in the near future, they will be satisfied. Late-night radio talk shows can be, for the lonely and alienated, a minor source of pleasure, removing momentarily the pain of isolation from other human beings; but the need for human contact remains intact, only briefly deflected and forgotten. The hope that better tranquilizers will be found, for instance without harmful side-effects, energizes further and unending research; and individuals change drugs, hoping to find a better one.

The relative success of technology in meeting some very old needs and its apparent success in meeting some new ones is enough to guarantee its future, a future which no number of disasters could nullify, nor any number of exposés of technologically exacerbated needs do away with. The last man and the last woman, survivors of the great nuclear holocaust, will fashion themselves a pick and shovel with which to dig themselves out of their deep cave; and the technological process will start all over again. Since technology will survive as long as man survives—and the logic of technology has historically proved to be more and better—the only rational possibility open to man is understanding and control. Understanding is necessary in order that the motivation behind technology as a whole, and different technologies in particular, can be grasped. Only if they are grasped will it be possible to devise means, the less hazardous the better, to satisfy the needs from which they spring. Control is necessary because of the simple and inescapable perception that technology can be hazardous to life, limb and mind. The technological price of satisfying one need can be that of subverting the possibility of realizing other needs.

The harm done by technology is both direct and indirect. Nuclear weapons are directly harmful; they are designed to kill. The harm of improved medical technology is ordinarily indirect, most prominently, at the present historical moment in the reduction of deathrates and the subsequent increase in population growth rates. Yet both technological developments have one thing in common. They can harm those who wield them as much as those they are meant (in the case of nuclear weapons) to destroy or (in the case of medical technology) to save. Even so, because harm can be either direct or indirect, technologies must be evaluated in different ways. The price of a failure in underdeveloped nations to develop new agricultural methods, to reduce infant diseases and to institute good medical programs, will be the predictable death of many human beings. That those saved, or their children, may die eventually of the effects of overpopulation is, for all that, to be part of a different moral drama than that represented by the proliferation of nuclear weapons. Analogously, while we know that air pollutants can kill, we also know that it is very damaging to a sense of security to endanger the jobs of workers by enforcing high standards of pure air. Survival for a worker is a job, not clean air, however much dirty air may kill him in the long run. A worker is done no service if we neglect the former while fulminating against the latter.

The control and proper use of technology encompasses all of the foregoing points. Technology will not be controlled if, in the process, the needs from which it springs are ignored. Where it springs from needs falsely realized—which are, nonetheless, quite real to those whose needs they are—substitute and genuine objects of satisfaction must be provided if the problem is that of false or dangerous objects. If the problem is that of needs which would not be so exigent but for technology, then the technological system must be scaled down to do away with those intensities. Human beings will always be anxious about their own preservation. They need not (conceivably, that is) be anxious about their preservation in the face of nuclear weapons or of competitive industrial living—those could they be done away with.

Most technologies are not directly destructive. The major issue with most is their power for indirect harm, either that harm which results from excessiveness (the sheer quantity, for instance, of automobiles) or from harmful by-products (an upsetting of the balance of nature). It is at that point that the satisfaction of one need can jeopardize the possibility of satisfying others. How is a balance to be achieved? It will not be achieved by a simple-minded "technological assessment." If such assessment is understood only to be an analysis of the consequences of different technologies—and that is all it is usually understood to be—it will not take us very far. Eventually we will have to ask the question: Just what is *good* for human beings? Which consequences are acceptable and which are not?

The debate over the SST is illustrative. The main and most telling arguments against American development of the SST turned on its potential environmental hazards. Yet these were effective arguments only because the proponents could not make out a strong enough case for the economic or preservative benefits of the SST; some people would be thrown out of work if the project did not go forward, but most would not. Quite the opposite happened in England and France, where a successful economic-political argument was made. In neither case, however, was there any basic consideration of what human beings need, or of how or when they need it. Moreover, if it is the case that SSTs pose environmental hazards, those manufactured abroad pose as many dangers as those manufactured domestically. The case for international control—and perhaps prohibition—is even stronger than the case for national control. There is little talk of introducing such controls. The United States, for one, shows little interest in endangering its relationships with its economic and nuclear partners in France and Great Britain by fighting their development of the Concorde. The direct harm to the environment which the Concorde may produce is, in the end, seen to be preferable to the envisioned indirect injury to the technological-economic-military network represented in the American, French and British alliance.

That such choices are made is indicative of another facet of technology: its cumulative power. My classification of technologies does not take account of the fact that different technologies feed upon and strengthen each other. In terms of causal connections it is evident why "corrective technologies" can so easily give rise to "enhancement technologies." But it is no less evident why quite disparate technologies can strengthen and come to depend on each other. All forms of technologies, for instance, require facilitating technologies designed to make other forms of technology function efficiently and effectively. Effective weather control requires the service of computers, and efficient corrective surgery requires tool-making machines to produce surgical devices. In turn, facilitating technologies produce jobs, and jobs produce income to buy the food made possible by adaptative technologies. The larger populations produced by more effective adaptative and corrective technologies require, in their turn, even more facilitating technologies if they are to survive. This is only to note that technology is in its cultural setting an entire system, with different bits and pieces serving, and compounding in size and scope, the system as a whole. The cumulative effects of the system can, in a very real sense, far exceed the sum total of its parts. Not only is it difficult to reduce or eliminate one technology without affecting most of the others, but it is no less difficult, once the system is in full operation, to adjust the human needs and desires which the system, wisely or unwisely, is haphazardly designed to serve. That is a social reality principle of the most stubborn and intractable kind.

I have taken us, I am afraid, into a swamp, or to use a more apt image, into a sticky cobweb. The more we squirm to get out of the cobweb, the more of it we find clinging to us. Unlike the spider, however, who can make an intricate web and never get caught, man has not found a way to create technological webs which do not trap and on occasion destroy him. Yet man needs his web no less than the spider needs his. So far I have done no more than urge that we look at the different strands in the technological web, asking what purposes they are meant to serve. I have also urged that it is helpful to distinguish between true needs and false satisfaction of them, and between direct and indirect harm. Implicit in that discussion was a further distinction between the motivation behind various technologies and the actual effects of those technologies. The whole discussion presumes not only that man needs technology, but that man is by nature technological. If we reverse that premise and ask whether technology is by nature human the answer is less clear. Obviously technology is a human artifact; it can come into existence in no way other than by human design and inventiveness. Just as obviously technology can destroy its maker. But in this respect, technology only apes human behavior, for human beings can destroy each other even without technology, though less efficiently and rapidly.

Technology, then, is nothing but an extension of man, displaying all the tendencies and characteristics man himself displays. The problems posed by technology are, for that very reason, reducible to the problem of what man is to do with and about himself. Technology complicates that question—complicates it immensely; it does not essentially change it. In that respect, Freud is not to be faulted for minimizing the dangers of technology. While it is true that technology seems to take on a life of its own, and to introduce a logic and set of requirements which are also its own, I do not believe it actually does so in any ultimate sense. For it is human beings who choose in the very first instance to live technologically, to take on its logic and its requirements (because to some extent they must). Still, because it is their needs and desires they are trying to satisfy when they do so, they have at least the possibility of asking how those needs and desires can *best* be satisfied. That *any* satisfaction will require *some* technology must be seen as a given. But how much, and in what way, and when, is not a given, except within very narrow limits of adaptation and protection.

Historically, preservation technology takes precedence. Without adaptative, protective and corrective technologies (subsumed under preservation technology), none of the others, or man himself, would be possible. That is still the case and why, incidentally, it is so graceless for advanced societies to warn the sick and the starving in underdeveloped countries of the hazards of technological progress. For them, and quite rightly, sufficient to the day is the evil thereof. To say this is only to reiterate the importance of distinguishing between and among the needs different technologies are meant to serve, and the equal importance of thinking about technology historically and contextually, in time and place.

Once we do that we will be better placed to understand where attention ought to be concentrated. Precisely because it is so fundamental, preservation technology requires the most delicate evaluation. On the whole, it is asking too much of man to curb his desire to survive. At best, we may only be able to point out that certain forms of preservation technology are actually counterproductive; they reduce rather than enhance the possibility of adaptation, protection and correction. There is some hope that, with enough education, limits will be seen and acted upon. Improvement technology, by contrast, is open to considerably more skepticism. Although there is no denying that human beings would like to go beyond what are now the established limits of possibility—the desire is evidently there—it has yet to be shown that there is an actual need to do so, a need which, if not satisfied, would somehow keep man from realizing himself. "Enhancing" technologies should be subjected to particular scrutiny, if only on the grounds of whether the proffered enhancements would actually be that. They presume we know what a "better" human being would be—and it may be doubted that anyone really understands that. The case for "decorative" technologies is even weaker; it is hard to see what of genuine value would be lost if most of them were done away with. "Aesthetic" technologies present a more mixed picture. Human beings appear to need beauty; and beauty is taken to improve man. At the very least, it should be legitimate to ask just how much technology the uncovering of beauty requires. It is considerably less than the constant development and proliferation of stereo systems would imply. Other generations managed to give music a central place in culture, without the help of either electricity or steel.

"Implementation" technology, heavily dependent upon the existence of all the other technologies, will be coped with and controlled only to the extent that the others are. While it is useful to ask whether any particular implementation technology is actually needed, or whether benign substitutes could be found for those which are hazardous, it is more important to see their parasitic relationship to the technological system as a whole. Scale down the system and they will be automatically scaled down as well. As for "destruction" technology, let him who will speak up for it—and be prepared to bear the consequences. There is no reason for the rest of us to be so prepared. "Compensatory" technologies will endure as long as compensation is needed; and no end is in sight.

## NOTES

1. Sigmund Freud, *Civilization and Its Discontents,* trans. Joan Riviere (Garden City, N.Y.: Doubleday, 1958), p. 38.

# CHAPTER 4

# *The Uses and Abuses of Survival*

In any evaluation of contemporary technology, two social realities must have a prominent place. One is the pressure for survival, at the historical roots of the development of technology, but no less powerful now in its force than in earlier generations. The successes of technology have by no means reduced a preoccupation with survival; in many respects, particularly the psychological, the preoccupation has become an obsession. The second social reality is that of individualism, which has fueled the development of technology, strengthened its ideological base, and exacerbated almost beyond correction the tendency of people to seek in technology a miraculous cure for their private miseries.

It is no accident that the cry for survival—implying in its ethic a communal control of communal destiny at the price of individual welfare where necessary—has reached a peak of intensity at just that moment when individualism can point to spectacular triumphs. In their private lives, individuals can increasingly do what they please, devising their own personal, sexual and religious modes to suit their fancies or visions. Yet they have become semiconsciously aware that when each seeks his own, and only that, the survival of the whole is threatened; and in ways I will develop, that their own sense of psychological survival as individuals is no less threatened. At no time in human history has it become more possible to test the validity of Freud's assumption that there is an irreconcilable conflict between the needs of the individual and the requirements of civilization. For at no time in history have individuals been so free to pursue their private directions without hindrance; and yet, if we listen to those who preach the gospel of survival, at no time has mankind as a whole been so endangered. Both the logic of individualism and the logic of survival are being pushed to a *reductio ad absurdum*.

No biological axiom is more firmly established than the instinct of survival.[1] Every species seeks to live and to perpetuate itself. The history of existing species can be recounted as a drama of adaptation and survival. The history of extinct species, by contrast, can be told as a tale of adaptative failure and consequent demise. Tolstoy's observation, in the opening lines of *Anna Karenina,* that all happy families are alike and all unhappy families different, seems no less true of species. Those which survive share a common trait: they are able to adapt to changing circumstances. Those which die out do so in a variety of ways, some rapidly and spectacularly, and others by slow attrition. The history of the human species is of a piece with the history of other successful species. Human life goes on because, over tens of thousands of years, it has developed the biological capacity to adapt itself in a phenomenal way to almost every earthly environment.

There are of course some distinctive notes in human survival. The most important is the fact that human beings, through the evolution of a superior brain, have been able to take conscious, intelligent control of their survival. Where other species have been dependent upon the luck of an advantageous draw in the genetic and environmental lottery, the human species—which must have started with a good draw in the first place—has, by means of wit and culture, been able to parlay its bet in a way impossible for any other species. We have not so far gone the way of the passenger pigeon or the dinosaur, both lovely creatures, but to their misfortune rather stupid.

The relationship between technology and survival was sketched in the previous chapter. Technology has provided essential tools for preservation and adaptation, yet excessive population growth—a consequence of technology—poses direct dangers to survival. It is quite possible to conceive of a world in which there are too many people to be supported by the available psychological or environmental resources. One has only to make certain projections from the figures on existing growth and consumption rates to come up with some disturbing possibilities. Add to this the recollection of what happened to individual tribes and groups in the past, decimated or wiped out by starvation, war or disease, and it is easy to imagine something similar happening to the whole species. Just this kind of imaginative speculation has been a powerful impetus behind a concern for world population growth rates, perhaps the first instance after atomic warfare where a worldwide concern with a by-product of technology was manifested. In terms of force and potency, a concern for survival—global and national—has overshadowed the myriad other arguments for population limitation, those which bear on economic development or the maintenance of a decent quality of life. The notion of extinction, utter extinction, is the most unbearable thought of all.

How should we understand the drive for survival in human beings? And how, as both a drive and a value, should it be approached and weighed in the context of technology? At the outset, some curiosities should be noted, which taken together point to deep mysteries. Unlike other animals, human beings are consciously able to kill themselves by suicide; some people choose to die. Unlike other animals, humans are able to give up their lives in the service of ideas, values and ideologies. Unlike other animals, human beings are not satisfied with mere adaptation and survival; they seem constitutionally predisposed to want more than life, and in general just to want more of everything. This is another way of saying that human beings want to be happy, a trait which leads them into religious and philosophical quests, the building of cultural and political systems, and an unabated pursuit of scientific knowledge and technological application.

The sum total of these species-specific traits suggests that the usual biological models and scientific laws relevant to nonhuman living species will have their drawbacks; and in practice that seems to be the case. Models which work with ants do not work well when extrapolated to human beings. If nothing else, it is an offense to good biological order that while all species, including the human, want to survive, only the human species sets out consciously on occasion to kill its own kind in technologically ingenious ways. Freud's emphasis on war was not misplaced. It is a positive biological outrage that one group of human beings will kill another in the name of the survival of all.

The central paradox of human survival is part of the paradox of human life. Why is it that while human beings, as individuals, groups and species, want desperately to survive, they will not settle for *mere* survival? Why is it that human beings are able to treat survival as both the highest value and the lowest value? Why, for the sake of goals other than survival, will they run the most extravagant gambles with survival? Why, when they increasingly know that unchecked population growth or proliferation of nuclear weapons or pollution of the atmosphere could lead to disaster or extinction, do they nonetheless give those hazards a low social priority?

It is not very illuminating to urge, as do some biologists and ecologists, that human beings are just stupid and shortsighted in this respect, perversely refusing to consider the folly of their ways. Indeed there is folly, if one wants to center on the need for survival as the sole important biological value. But the evidence of human thought and behavior suggests that human beings value many goods, of which survival is only one. Arguments which would work with animals, whose aspirations are limited to survival, will not work with human beings, who insist upon something more. Nor, to use a more sophisticated model, will it do simply to say that, while man's unique genetic makeup was invaluable in enabling him to survive in the past, it must now be modified and subdued in order that it will continue serving him in the future. This is the overadaptation model, which compliments man for his good beginnings but then criticizes him for not knowing when he has enough of a good thing. But that also misses a critical point about man, that he seems compelled to act for ends other than optimal adaptation and a satisfactory mode of survival.

The point of these remarks is to see if, at the outset, the problem of human survival can be set in some illuminating context. It is not altogether enlightening to be told that all sorts of drastic steps must be taken if the species is to survive. That is probably true, but it is too crude a truth to be helpful. It presupposes the aspiration level of an insect, who cares nothing about anything but survival. Nor is it illuminating to be told that man has overadapted. That may also be true, but it presupposes that human beings ask nothing other than optimal adaptation. In short, we are not likely to understand human attitudes toward survival, much less influence behavior in ways which enhance the possibility of survival unless we understand the exceedingly complex human response to the problem of survival in relationship to all of the other ends human beings seek.

More important, if there is a need for an ethic of survival in relationship to the uses of technology, the possibility of genetic engineering and excessive population growth rates, this ethic must be constructed from richer ingredients than the observation, however true, that population growth or other technological dangers pose a dire threat to the survival of the human race. It is doubtful that humans will be responsive to that kind of truth; and in fact they have not been. Moreover, I would want to argue that, in order to remain human, they should *not* be all that responsive. Or better, they should be responsive only to those survival arguments which manage to integrate the need for survival with a whole range of other human needs, some of which would risk survival for the achievement of higher values.

A beginning can be made toward this integration by noting some of the uses and abuses of the concept of survival. Historically, the uses have been more evident than the abuses. Among the uses are those of a fundamental perception of a biological reality principle: unless one exists, everything else is in vain. That is why survival, the desire to live, is so potent a force, and why the right to life is such a basic part of any reasonably enlightened social, political and legal system. Politically, particularly in time of war, national survival has been a potent force for mobilization of community effort, transcendence of self-interest, and creation of patriotic esprit. For individuals, the desire to ensure the survival of offspring has been the source of great and selfless sacrifice and the voluntary acceptance of obligations to future generations. Within the private self, a will to live, to survive at all costs, has literally kept people alive, staving off a despair which would otherwise have been totally destructive.

That individuals, tribes, communities and nations have committed so much will, energy and intelligence to survival has meant that they have survived, and their descendants are present to tell the tale. Nothing is so powerful a motive force, for self or society, as the threat of annihilation, nothing so energizing as the necessity to live. Without life, all else is in vain. Leaving aside the question of whether we need more enlightened attitudes toward suicide in our society, which we may, it is still not for nothing that suicide has been looked upon with abhorrence, whether from a religious or a psychological perspective. It seems to violate the most fundamental of human drives, and has always required a special explanation or justification.

The value of survival could not be so readily abused were it not for its evocative power.[2] But abused it has been. In the name of survival, all manner of social and political evils have been committed against the rights of individuals, including the right to life. The purported threat of Communist domination has for over two decades fueled the drive of militarists for ever-larger defense budgets, no matter what the cost to other social needs. During World War II, native Japanese-Americans were herded, without due process of law, into detention camps. This policy was later upheld by the Supreme Court in *Korematsu v. United States* (1944) in the general context that a threat to national security can justify acts otherwise blatantly unjustifiable. The survival of the Aryan race was one of the official legitimations of Nazism. Under the banner of survival, the government of South Africa imposes a ruthless *apartheid,* heedless of the most elementary human rights. The

Vietnamese war has seen one of the greatest of the many absurdities tolerated in the name of survival: the destruction of villages in order to save them.

But it is not only in a political setting that survival has been evoked as a final and unarguable value. The main rationale B. F. Skinner offers in *Beyond Freedom and Dignity* for the controlled and conditioned society is the need for survival.[3] For Jacques Monod, in *Chance and Necessity,* survival requires that we overthrow almost every known religious, ethical and political system.[4] In genetics, the survival of the gene pool has been put forward as sufficient grounds for a forceful prohibition of bearers of offensive genetic traits from marrying and bearing children. Some have even suggested that we do the cause of survival no good by our misguided medical efforts to find means by which those suffering from such common genetically based diseases as diabetes can live a normal life, and thus procreate even more diabetics. In the field of population and environment, one can do no better than to cite Paul Ehrlich, whose works have shown a high dedication to survival, and in its holy name a willingness to contemplate governmentally enforced abortions and a denial of food to starving populations of nations which have not enacted population-control policies.

For all these reasons, it is possible to counterpoise over against the need for survival a "tyranny of survival." There seems to be no imaginable evil which some group is not willing to inflict on another for the sake of survival, no rights, liberties or dignities which it is not ready to suppress. It is easy, of course, to recognize the danger when survival is falsely and manipulatively invoked. Dictators never talk about their aggressions, but only about the need to defend the fatherland, to save it from destruction at the hands of its enemies. But my point goes deeper than that. It is directed even at a legitimate concern for survival, when that concern is allowed to reach an intensity which would ignore, suppress or destroy other fundamental human rights and values. The potential tyranny of survival as a value is that it is capable, if not treated sanely, of wiping out all other values. Survival can become an obsession and a disease, provoking a destructive singlemindedness that will stop at nothing.

We come here to the fundamental moral dilemma. If, both biologically and psychologically, the need for survival is basic to man, and if survival is the precondition for any and all human achievements, and if no other rights make much sense without the premise of a right to life—then how will it be possible to honor and act upon the need for survival without, in the process, destroying everything in human beings which makes them worthy of survival? To put it more strongly, if the price of survival is human degradation, then there is no moral reason why an effort should be made to ensure that survival. It would be the Pyrrhic victory to end all Pyrrhic victories. Yet it would be the defeat of all defeats if, because human beings could not properly manage their need to survive, they succeeded in not doing so. Either way, then, would represent a failure, and one can take one's pick about which failure would be worse, that of survival at the cost of everything decent in man or outright extinction.

Somehow we need to find better alternatives, if I may be allowed to understate the matter. We need to survive as races, groups, nations and as a species, but in a way which preserves a wide range of other human values, and in a way which is as sensitive about means as about ends. Control of technology and population limitation will be an essential means to survival of the species. Thus the problem is to find a way of living with and profiting from technology, and of controlling population growth, size and distribution which is as morally viable as it is pragmatically effective. A balance will have to be devised, of the most delicate kind. A number of steps are necessary, the first of which is to analyze the various types of supposed threats to survival. At the very least, we need to know which are real and which are imaginary, which are of the essence and which are fantasies. We also need to have a sense of those other values human beings prize, especially those for which they are willing to risk survival, even to give it up altogether. In sum, we need to know just what it is we are trying to balance, and what would count as a good balance.

A number of types of survival can be distinguished, the most important of which are survival of the species and survival of nations, cultures, groups (racial, ethnic and religious) and individuals. Survival of the species provides the prototype concept of survival. Taken literally, it can be understood to mean a continuation of human existence, specifying nothing about the number of those existing or the quality of their existence. In that sense, the species could survive if only a handful of fertile humans existed, much as the bison or the California condor exists, and even if the level of existence was that of a primitive tribe. If survival of the species alone is the goal, understood in a minimal sense, it is reasonable to suppose that nothing less than a global, all-encompassing catastrophe would suffice to bring about extinction. Nuclear warfare, together with a persistence of life-extinguishing levels of atmospheric radiation, might present that kind of threat.

It seems to me difficult, however, to imagine any other kind of catastrophe which would have a like effect. Pollution of the gene pool would take thousands of years, even if total pollution is conceivable in theory. Overpopulation would, well before human extinction, be a self-correcting phenomenon. People would die until a supportable number remained, a state which could be reached well before extinction became an imminent reality. To be sure, excessive population growth could conceivably bring about a worldwide nuclear war, as people and nations struggled for more space and resources. And I suppose it is possible, in a world of steel, concrete and carbon dioxide fumes, to imagine oxygen shortages. But those are the only circumstances in which it makes much practical sense to talk about the extinction of the species. To be more blunt, the spectre of total human extinction is a chimera, providing a poor base upon which to build a concern for the necessity to control technology. Disasters could happen, under some remote circumstances; but then any and all kinds of catastrophes are imaginable under some circumstances.

These remarks are not meant to dismiss survival as a concern, though. If the concept is understood in a wider, non-literal sense, it is serviceable and important. Let me stipulate that sense as the continuation of the human species at a level of health and subsistence which makes possible the development of culture and individual self-fulfillment. That definition of survival leaves open the question of the rate of population growth and the number of people the earth can sustain, as well as the level and kind of technology which can be borne. The optimum population of the earth, or the optimal level of technology, encompassing some optimal notion of survival, cannot be a fixed figure. Too much depends upon the kind and quality of life desired or lived. If the ideal model is the affluent American, that will produce a very different degree of magnitude of consumption and use of technology than if, say, the model of a healthy, middle-class Indian is used. The central question is not so much whether the human species will survive, but how it will survive. And this question cannot properly be answered without a consideration of what human beings actually need to live a life of dignity and fulfillment. At the least, with the possible exception of health care, it is certain that human life does not require for its fulfillment all the goods that the affluent American thinks he needs. On the contrary, quite apart from population considerations, our version of technological affluence carries an excessively high psychological and social price.

Survival also remains an important value if further extended to include the survival of nations, cultures, groups and individuals. Most people, one might guess, are actually more concerned about their own survival, the survival of their family and of the social group with which they most identify, than with survival of the species. They live in the macrocosmic world, but it is their microcosmic psychological and social world which counts. A threat to that latter world will be far more immediately unsettling than a threat to something so abstract and impersonal as the species. How many of us would be desperately concerned about the fate of the species if, along the way, our family, our descendants, our culture and our social groups were certain to disappear?

A distinction can be drawn between personal and social survival, on the one hand, and species survival, on the other. Personal-social survival refers not only to the survival of individuals, but to the survival of those groups which provide them with their culture, values and social identity. Since the rise of the nation-state and with the acceleration of worldwide nationalism, the nation has been added to older, traditional social groupings as a source of individual identity and as the primary economic unit supporting individual welfare. In the United States and elsewhere in the world, the survival of racial and other types of minority groups has recently taken on a new power. In developing nations, conflicts among religious and ethnic groups—each anxious about its survival and strength—are a major impediment to the development and implementation of coherent national social policies.

When survival is seen in this broader context, a variety of ambiguities can appear. The survival of a nation, for instance, can be understood in a number of senses. It can mean literal survival, often the focal point, real or imaginary, of debates over the amassing of weapons. It can mean political survival, that is, the survival of a particular set of political and legal institutions. It can also mean the survival of a particular way of life, as in references to the "American way of life." The emphasis there can fall on the preservation of certain cherished values, such as freedom or economic individualism, or on certain levels of economic well-being and employment opportunities. It can also mean survival as a world power, or less pretentiously, survival as a political and economic entity which commands the respect of others. If a willingness to go to war is taken as one test of a nation's perception of a threat to its survival, then it is evident from history that all of the above senses of national survival have at one time or another come into play.

Within nations, and often across national borders, the survival of religious, racial and ethnic groups has been a powerful motive in fostering or opposing population control policies. While the dynamic here requires considerably more research for a clear understanding of the forces at work, a few characteristics seem evident. For many individuals and groups, there is a far stronger sense of identity with their racial, ethnic or religious group than there is with the nation in which they happen to live or to have been born. This may happen either because, taken positively, these groups provide a stronger source of meaning and community bonding than national identities (particularly in pluralistic societies) or, negatively, because of a fear (often rational enough) that the majority populations in particular nations are hostile toward them.

The fear of genocide among blacks in the United States is a case in point in this latter context. Genocide can be understood as physical extermination in its most extreme form, or as stripping away from a minority group its culture, self-respect, political power, self-direction or distinctive identity. Both types of genocide have an appalling number of historical precedents, and a fear of socio-cultural extinction is often as strong as a fear of physical extinction. For fearful minority groups, whose heritage is that of persecution, refutation of the charge that physical genocide is intended by no means suffices to show that socio-cultural genocide is not at work. It is beside the point that socio-cultural genocide is not consciously intended. The same result can occur if the policies devised and imposed by the majority impose a greater burden on minorities than on themselves. A population policy designed by middle-class whites, seeking the two-child norm per couple, will inevitably seem a piece of cultural imperialism. Blacks will be quick to note that this norm imposes few burdens on upward-mobile, affluent whites, who are not likely to want more than two children anyway, regardless of population-control motives. They will no less quickly note that the values to be protected by the population policy, whether economic, political or environmental, are the values of that white majority, not their values.

So far I have tried to analyze the concept of survival and to note the various senses in which it is used and may be understood. Perhaps it can all be summed up in the following way. The power of the drive for survival draws on biological, psychological and social roots. As individuals, we fear death and extinction. That fear seems universal, attested to by scientific, literary, religious and philosophical evidence dating back to the beginning of human consciousness. But we also fear, perhaps no less strongly, the destruction of our psychological and social worlds of meaning and identity. A blow to the ego can be, in its perceived power, as strong as a blow to the body. A blow to our primary reference group—whether that group be national, racial, religious or ethnic—can be as threatening as a blow to the private self. For many or most of us, the private self cannot be sharply distinguished from the communal self which we share with others in our group. More people commit suicide because of a shock to their sense of self-worth and identity than because of dire physical illness. And human beings seem as willing to kill or be killed in defense of their social group as in defense of their individual life. We simply cannot understand human motivation or behavior if we do not understand both the pervasiveness of the drive for survival and the great variety of ways, individual and social, in which it manifests itself.

At the same time, however, we will no less fail to understand the place of survival in human life and history if we do not also observe the way it can be complemented by, or subordinated to, a wide range of other values. I have already contended that few human beings, at least in the long run, will settle for mere physical survival. Prisoners will risk their lives to escape from jail. Slaves are willing to die in attempts to be free of their masters. Colonial nations will revolt in the face of overwhelming odds to achieve their independence. Nations will go to war to preserve their sovereignty and some, however insane it appears to be, will build and store nuclear weapons in massive overkill quantities rather than be losers in the nuclear arms race.

With some ingenuity, I suppose it would be possible to develop a theory to show how all such acts, despite their appearance, are no more than an adumbration on the theme and drive of survival, particularly in its social and psychological variants. Anyone who wants to develop such a theory is welcome to do so. I find it equally plausible, however, to understand such acts by seeing in them a manifestation of human needs and aspirations which, if they do not necessarily transcend survival in their force and weight, can certainly be as strong in many cases. The American Revolution was not fought to preserve physical survival; colonial Americans were certainly in no danger of physical extinction or even in danger of serious poverty. That revolution was fought to advance the values of freedom, justice and self-determination. When Israelis say "never again" and assert their willingness to die rather than accept subjection, they are making clear their need for something more than survival, their freedom and dignity as a people. The Jewish case is particularly significant in another respect also. The lesson which many Jews believed they learned from the holocaust is that the most ineffective way to guarantee survival is to be passively willing to settle for survival. Unless one wants more than survival, and is willing to die for it, even survival will be taken away.[5]

A similar dynamic can be seen in the rise of black militancy in the United States. The perception of the militants is that the passivity of an older generation of blacks, a hangover from the years of slavery, is naïve and counterproductive. Far from assuring the survival of blacks, it makes them easy prey for racists, whether the motives of the latter are literal or cultural genocide. Both Jews and militant blacks, whatever their other differences, have perceived that their survival will always be in jeopardy as long as it is dependent upon the goodwill and toleration of others. And if this is true of survival, it is no less true of the protection of freedom and justice.

There are two general points which can be drawn from the variety of examples I have offered. The first is that it is possible to conceive of a variety of circumstances in which an excessive emphasis on survival, casting aside other human needs and values in the process, can severely jeopardize survival. Nuclear weapons, originally developed for self-defense, can be seen as the great symbol of that thesis. Almost as powerful a symbol can be found in excessive population growth rates, which stem, we should recall, from the powerful success which medicine has had in meeting the need to reduce deathrates, i.e., in meeting the demand of human beings to survive rather than to die. The second point is that a viable, human ethic of survival must encompass all the other values and goals human beings seek and prize. Otherwise, it will either be self-defeating or productive of a life which most people will not find worth living.

A third point also needs development, to complement the first two. Survival as a need and value will inevitably be perceived in different ways by different individuals and groups. While it is now possible to determine scientifically what people need in the way of nutrition to survive physically (e.g., caloric and protein intake) and from that knowledge to project the kind of agricultural and environmental resources necessary to sustain minimally adequate diets for given numbers of people, not much else is known with any certainty. Important research is being done on the effects of crowding on behavior, on the limits of environmental pollution, and on the extent of natural resources. But so far I think it not unfair to say that little scientific consensus has been reached on any of these problems, each one of which affects human survival. The most we know is that the earth cannot support an infinite number of people infinitely exploiting natural resources and infinitely polluting the environment. But that does not tell us much about what an optimal population size or growth would be or how much technology can be endured or what the limit is to the range of possible physical and psychological adaptation. One thing evident from debates about the limits of the earth's carrying power, and about optimal population size, is that the ideological values brought to bear on existing data are enormously diverse. There is little agreement on just what exactly is necessary for mere survival, on the one hand, and for survival with dignity, on the other.

The problem becomes even more complicated in the light of the great variety of habits, traditions and expectations which govern what various individuals and groups perceive as necessary for the kind of survival they can, so to speak, live with. The recent back-to-nature romanticism of some middle-class Americans is based on the premise that a happier life can be found in an existence close to the earth, spared the excesses of technology. Less-developed countries, more experienced in living close to the earth, are rarely subject to such romantic fits. High infant mortality rates and short life-expectancies do not encourage the view that a technology-free life is conducive to high levels of survival. The fact, however, that there exists a revolt against technology in many places in the developed countries, while at the same time the drive for a more advanced technology continues apace in areas which do not yet have it, suggests that a middle ground must be sought.

Both a deficiency and an excess in the power of survival as a motivating drive breed anxiety and discontent. Too low a survival level forces a constant confrontation with annihilation, an unbearable kind of burden. Yet too high a survival level seems to produce a very similar kind of response. In the former case, it is the fear of physical destruction which is paramount; in the latter it is that of psychological destruction. Technological societies have achieved physical security at the price of psychological and social insecurity. The combination of individualism, consumerism, the generation of unlimited material desires, competitiveness and anomie, together conspire to make psychological survival fragile and precarious. If this felt fragility had the effect of reducing the demands made on technology, a more rational balance might be found. Unfortunately, it seems to accelerate the demands, particularly in the development of drug and behavioral therapies—compensatory technologies—which do not strike at the roots of the insecurities but seek only to ameliorate the symptoms.

A parallel phenomenon appears to be at work on the national level. Every one of the nuclear powers in the world displays anxiety about the adequacy of its power. The result is a constant spiral of nuclear arms and their delivery systems, with each nuclear nation subject to constant internal pressures for the development of ever more sophisticated methods of defense, deterrence and first-strike capability. The net result is not a greater sense of security and survival, but less. One could hardly seek a better illustration of the tyranny of survival, a tyranny, it is clear, which can dominate the rich as powerfully as it dominates the poor. For that matter, the very poor are often fatalistic, not expecting or hoping to survive; the rich do not give up so easily.

The major difference between the rich and the poor, however, is that the culture of technological, affluent cultures has established a historically unprecedented survival level. Its point of departure is a very high base-line, one which appears to socialize individuals in a way which makes them unable to conceive of an acceptable survival which does not include all the comforts, power and symbols of advanced technologies. These technologies in the nature of the case seem to need constant improvement even to keep pace with the ever more expansive, seemingly unlimited survival demands they have historically generated.

My point here is a variant on the familiar thesis that the consumption levels and style of life to be found in affluent, technological societies pose a global danger to the preservation of natural resources and to the environment, and a psychosocial hazard in the living of daily life. I believe that to be a reasonable position. But I am also asserting that the motive force behind this hazardous phenomenon is that same drive for survival which dominates the poor, nontechnological societies. The difference is that the emphasis in the rich countries shifts from the physical to the psychological level. At that level, two things appear to happen. The first is that the survival demands become increasingly difficult to fulfill, mainly because so much more, an infinity more, is asked; frustration and dissatisfaction are generated. The second is that the means chosen to fulfill the survival drive tend to be increasingly counterproductive, requiring a constant spiral of compensatory technologies. Just as the survival potential of the nuclear powers is probably less, in the end, than that of the militarily weaker nations (in the sense that they are liable to a more total destruction) so the psychological survival power of individuals in technological societies is probably less than that of individuals in less-developed countries. Even at the physical level, the urban dweller is in a curiously weaker position than many inhabitants of poor countries; he cannot raise his own food.

As I hope the foregoing discussion will have made clear, the relationship between survival and an optimal use of technology and population growth is complicated by a number of shifting variables. The need for survival is modified by the need to realize other values as well, notably freedom, justice and a sense of dignity and worth. The meaning of survival, once one moves beyond the level of bare subsistence, will be subject to a variety of different national, group and individual interpretations, primarily because survival will usually be interpreted in terms of desired standards of living and the preservation of values seen as integral to a satisfactory self-identity. The problem which remains is to see if it is possible to set forth some general standards concerning the use of survival as a value.

The first requirement is that a way be found to respond to the need for survival without, at the same time, allowing that need to become a tyranny. The tyranny can result either because of a panic in the face of a genuine threat to survival, because survival is invoked for self-interested or totalitarian political purposes, or because of an unnecessarily or unrealistically high standard of acceptable survival. Perhaps it is possible to do no more in the face of the last two possibilities than to be aware of their potential force, and by political and cultural debate to neutralize or overcome their baneful effect. The panic which can result from a real threat to survival will be more difficult to cope with, a panic which can lead to draconian measures in the name of self-preservation. At that point, the question must be faced whether there can be such a thing as too high a price to pay for survival. I believe there can be, particularly when the proposed price would involve the wholesale killing of the weak and innocent, the sacrifice to an extreme degree of the values and traditions which give people their sense of meaning and identity, and the bequeathing to future generations of a condition of life which would be degrading and dehumanizing. The price would be too high when the evil of the means chosen would be such as to create an intolerable life both for the winners and for the losers. While it might be possible to conceive of individuals willing to have their lives sacrificed for the sake of group survival, it becomes more difficult to imagine whole groups being willing to make such a sacrifice. And there is a very serious moral question whether that kind of sacrifice should ever be asked for or accepted, even on a voluntary basis.

The worst problems, however, are likely to arise not at the extremes but at some intermediary points. Certainly it would be reasonable and ethical to ask an entire population to give up to some extent some of its cherished values in the name of survival. Which ones and to what extent would be for them to determine. A minimal and first requirement, though, if that kind of demand is to be made, would be mutual agreement. If the values to be sacrificed were central and essential to the population's sense of its own dignity, I believe near unanimity would ethically be required, with the burden equally distributed among the entire population. This is not a case where a plurality of votes or opinion should reign. It is much too easy to imagine a powerful majority happy to purchase its own survival by taking what it needs from a weaker minority. Nothing less than an almost total consensus should be required if the means of survival require the taking of severe action. It would be well within the rights of a minority to jeopardize the survival of all if they correctly perceived that a majority intended to kill them to save its own skin. It would be no less the right of a minority to refuse to cooperate in a program designed to preserve a certain level of survival if that program was designed to require inequitable sacrifices from the minority.

The second requirement is that a way be found to determine (*a*) when survival should first be taken into consideration, and (*b*) when it should be allowed to become a social priority. My phrasing of the problem presumes the utility of understanding survival as a continuum, requiring points of demarcation in order that the right action may be taken at the right time. Survival should first be taken into consideration when there is evidence that the environmental, social and psychological costs of technological and population growth could, if unchecked, lead to a serious endangering of basic physical needs, a threatening of cherished national or group values, and a subverting of psychological security and identity. Note that I used the phrase "taken into consideration." The point of that wording is to suggest that too precipitate an action, when the hazard is still distant and problematic, may—given the possible tyranny of survival—do more harm than good. At the early stages, the wisest course may simply be that of public education about the possible consequences of present behavior and attitudes.

But let us assume that the stage of a dark cloud on some distant horizon has been passed, and the evidence is good that serious deterioration has already set in. At what point in the deterioration should survival become a priority? Observe that

I said *a* priority; it should never become *the* priority if that means the sacrifice of all other values. But there are surely conditions under which it could become a priority, and a very high one. The most important of those conditions would be the existence of evidence that irreversibility was beginning to set in, making it increasingly impossible to return to the original conditions. That situation, combined with visible evidence of serious present deterioration—for instance, an urgent need to develop compensatory technologies—would warrant a focus on survival; for that is just what would be at stake.

The third requirement is that a way be found to take account of the obligations owed to future generations. It is possible to conceive of present forms of behavior which, while posing only minor hazards and inconvenience on the living, could in their cumulative effect over a long period of time pose a serious danger to those yet unborn. The test of irreversibility is again useful. Are present actions such that their consequences would not only be harmful to those not yet born, but also of a kind which could not be reversed by them? Destruction of irreplaceable natural resources, permanent contamination of the environment, thoughtless genetic engineering, and the bequest of an excessively large number of people would be apt illustrations of this possibility.

The most vexing problem here bears on the degree of sacrifice which can be demanded of the living in the name of the survival of the unborn. That demand cannot, I believe, encompass the destruction of basic liberties of the living or a deprivation of those physical necessities of life required for their survival. But it could well encompass a radical shift in the style of living, in levels of consumption, and in the number of children brought into the world. The sticking point here would be whether shifts of this kind should be made involuntary. In the end, I think not, if only because the rights of the living, not to mention their political power, will transcend the rights of those yet to be born. But a vigorous educational campaign, even of a propagandistic nature, would not be out of order.

In the end, however, any recommendations about steps to be taken to protect the possibility of survival will seem particularly banal. If the hazards to survival are in the distant future—even a decade off—it is unlikely that many will be willing to give up their usual inclinations and day-to-day desires to protect their future lives. And if the hazards are immediate, the tyranny of survival, not conducive to rationality or a careful balancing of values, will take over; the most likely result is chaos, with the powerful imposing their will at the expense of the weak. Life in a technological society will exaggerate these tendencies to an extreme degree. The individualism of technological life, where the goods of technology are seen primarily in terms of private well-being, works against voluntary individual action taken for the common good. It is not likely to extend beyond such trivial steps as a greater willingness to pick up stray pieces of trash on the street in the name of environmental protection.

The massiveness of technological society reinforces resistance to individual self-abnegation. When the net result of private action is statistically miniscule, and evidently so to the individual, there seems all the less reason to do anything at all. Thus if something is to be done for the sake of survival, it will almost certainly have to be done by government, through law and naked power. In Freudian terms, technological societies have the effect of dampening if not destroying that aspect of the cultural super-ego which would lead a community toward altruism and greater unity. Size alone can have this effect, but size combined with the individualism endemic to advanced technological societies guarantees it. Moreover, when the base-line of survival for individuals reaches the high point it does in technological society, it will become increasingly difficult for individuals to even imagine how they could personally survive in a setting which required that they reduce their own survival demands. Lacking such an imagination, they are likely to do nothing. That has been the history of the nuclear arms race and, more recently, of attempts to deal with environmental problems. That governments can, for their part, duplicate the self-interested, protective reaction of individuals in their response to survival needs is self-evident. But that governments, once aroused, can also respond with viciousness to meet the needs is no less self-evident. When the tyranny of survival takes hold of individuals, they can wreak considerable harm to those around them, but only within a relatively small circle. Governments, by contrast, can impose massive harms on a community; and the technology at their disposal to do so will far exceed in power and sophistication that available to individuals.

Survival needs arise in times of scarcity. In earlier generations, when the perceived and experienced base-line of survival was lower, scarcity could be defined in modest, realistic terms. In our time, shaped by technology, scarcity is defined in a grandiose way, and in a way far more resistant to a scaling downward than was the case in the earlier phases of modern technology; panic at imagined personal loss will more quickly arise. The way Freud posed the dilemma between the demands of civilization and the demands of the individual remains pertinent. But he saw only dimly the way in which technology can change—by raising the psychological ante—the demand of the individual for survival over and against civilization, refusing any longer to make those sacrifices (never ungrudging in any event) necessary for communal survival. Nor did he see clearly that technological life, simply because of the individual survival demands it instills by virtue of its individualism, poses hazards to the realization of Eros as great as that of the destructiveness of war

## NOTES

1. I have profited much from discussing "survival" with Martin P. Golding, and from his monograph "Obligations to Future Generations," *Maoist,* Vol. 56 (January 1972), pp. 85–99.
2. Richard Neuhaus has many usefully biting things to say about the abuses of "survival" in *In Defense of People* (New York: Macmillan, 1971).
3. B. F. Skinner, *Beyond Freedom and Dignity* (New York: Knopf, 1971).
4. Jacques Monod, *Chance and Necessity,* trans. Austryn Wainhouse (New York: Knopf, 1971).
5. Martin Gottfried, commenting on the efforts of the director of a New York theater group to weather a financial crisis, makes a similar point about his efforts: "He has for so long faced animosity and political pressure . . . that he has learned to be conservative, to appease all critics and to try to please everyone. But by following this policy of survival, he has only intensified the insecurity that has plagued this theater from its birth, and created that most meaningful of all theaters, the frightened one. . . . Nor has that strategy worked in any pragmatic sense . . ." (*New York Times,* September 5, 1971, p. D–3).

# CHAPTER 5

# *The Tyranny of Individualism*

IN A TECHNOLOGICAL SOCIETY, the tyranny of survival could not flourish but for a more basic tyranny of individualism, escalating any threat to the private, self-determining self into a threat to life itself. The self-deception this requires is enormous, but well underway. We may indeed be in danger of eventual annihilation by a depletion of resources, polluted air and dirty water; it could come to pass. But before that happens we are more likely to institute a new reign of tyranny in the name of personal ego survival, bringing obliteration technology to bear on all those human beings who, as power and educated sophistication will decide, are perceived as the source of our misery. This is to say two things at the same time. There is in the hazards of technology a profound danger to the human future, whether through nuclear, environmental or demographic destruction. But the ultimate problem does not lie with technology as such, which has (as it has demonstrated) as much potential for good as for evil. Instead, it lies in the combination of technology and individualism, where technology finds its motive power (and its financial backing) in its purported services to individual well-being, providing the possibility of a personal "liberation" from the ills of life, whether physical, psychological or social.

The most penetrating critic of contemporary individualism is the sociologist Philip Rieff. Better than anyone he has seen the way in which that individualism not only dissolves all culture—and thus all limits to human behavior—but also that, mixed with radical technology, sets the stage for an unprecedented barbarism. If we are to understand the central paradox of our historical moment, the confluence of an obsessive concern with individual freedom and with species and group survival, we can do no better than to look to Rieff's account of the cultural and psychological setting of present-day individualism.

Anyone who can say, with what seems to be but is not a straight face, that mankind is now ready to move beyond "the outmoded categories of hope and despair" deserves our attention.[1] In *Freud: The Mind of the Moralist,* Philip Rieff established the vantage point which eventually led him to make that statement. In that book the aspect of Freud singled out for attention is "the mind of Freud, not the man or the movement he founded, as it derives lessons on the right conduct of life from the misery of living it."[2] In that sense, Freud was a "moralist," but unlike the run of moralists, the Freud of Rieff's portrait is that of a moralist without a message.

The very lack of a message gives Freud his severity as a moralist:

> Western culture in the twentieth century has produced no other equally severe, equally able to forego ideal convictions and analytic simplicities in the quest for the ordered life. Freud was not gulled by any of the false hopes or ego satisfactions in which all of us are educated; somehow he re-educated himself to avoid them and thereafter did not

> invent new hopes. . . . If he has achieved a kind of secular sainthood, if he can be used by the practical as well as invoked by the dedicated, it is because he taught us in a unique and subtle way how to grow unsentimental about ourselves. . . . Freud's own unsentimental attitude toward himself, as the first out-patient of the hospital culture in which we live, is an attitude desired by every thoughtful person nowadays, not as an end in itself but to assure that person that he is himself alone and not merely acting out various roles—now physician, now patient, now curious bystander—in the unending series of accidents and emergencies that constitute life, both private and public.[3]

The Freud whom Rieff lays before us is a secular saint because he prepared the way for the type of man who could survive his own epoch. Freud opened the way for a new revelation, by killing all revelation and teaching man how to live without it: "once again, history has produced a type specially adapted to endure his own period: the trained egoist, the private man, who turns away from the arena of public failure to reexamine himself and his own emotions. A new discipline was needed to fit this introversion of interest, and Freudian psychology, with its ingenious interpretations of politics, religion, and culture in terms of the inner life of the individual and his immediate family experiences, exactly filled the bill."[4] One's eye is arrested by that passage; it was written during the late 1950s, when the common wisdom was that rationalistic individualism had failed, that nothing less than a rebirth of human community would save man. "There is a decided weakening of faith," Robert Nisbet could write in an influential book published in the early 50s, "in the inherent stability of the individual and in the psychological and moral benefits of social impersonality. Impersonality, moral neutrality, individualism, and mechanism have become, in recent decades, terms to describe pathological conditions of society."[5] Who was the prophet here? When we look at the present scene, our eye can find evidence to support both views. There is at once a visible quest for community, heightened by the concern for survival, and a visible quest for a heightened individualism, at war with community demands and impositions.

The twin themes of individuality and community run through Rieff's analysis of Freud but, while he emphasized Freud's belief in the unifying drive of Eros, the more important element is the drive of the individual to break the bonds of community. Freud's "psychology," he wrote, "controverts the organicist idea that individuals realize their personal value through the *polis,* the church or holy community, or the state. . . . Freud's own sentiments were formed by an ascendant doctrine of individuality, according to which 'society' meant the sacrifice of individuality, not, as in earlier notions of organic community, its fulfillment."[6] Accordingly, in Rieff's view, when Freud coupled renunciation and good conduct, which creates the permanent tension between the individual and society, he inverted the traditional concern with morality: "moral ideas are named as the *problem* of life rather than as the basis for a solution to the problem. The Freudian dislike of social idealism, as a way of symbolizing and masking power, emerges in a powerful suspicion of sacrifice as sickness."[7]

One can only be suspicious of moral values and of political idealism if, at the summit of one's mountain, stands the figure of the untrammeled individual. Moral codes make demands upon people; political reform requires that something be given up for the sake of the new order to be achieved. The value of psychoanalysis, in line with Rieff's portrait of Freud as moralist, is that it makes possible a morality dedicated to the promotion of the private self: "Psychoanalysis is the doctrine of the private man defending himself against public encroachment. He cultivates the private life and its pleasures, and if he does take part in public affairs it is for consciously private motives."[8] The private life is unalterably opposed to communal life; they go their separate ways. When they impinge upon each other, as they must, then the bargaining begins. "Psychoanalysis," Rieff claims, "exposes the triviality of much contemporary agonizing over the conditions of social organization that encourage conformity. It undercuts the whole problem of the freedom of the individual in any society, emphasizing instead the theme of the anti-political individual seeking his self-protection in a context as far from the communal as possible. Psychoanalysis, by its turn to the inner life, puts fundamentally into question the received criteria for discriminating one regime from another, let alone for determining the best regime."[9]

Curiosity must be aroused at this point. Does Freud stand for anything at all? Rieff says that "Freud's is not a therapy of belief but one which instructs how to live without belief."[10] Yet must one not, even to live without belief, believe that at least unbelief is possible? To go through life without a belief in anything would, in itself, be a psychological miracle. For one must believe that all believers are wrong, and this stance can be adopted only on the basis of some utterly different perspective, and one, whatever it is, which would normally be called a belief. The man who can say that all believers are wrong knows more than the rest of us. How can one get in that enviable position? Rieff himself draws back from attributing an outright nihilism to Freud; he has, after all, presented him as a moralist. Something, then, must constitute the core of Freud's belief, some value must be present. Freud's commitment to reason, his god Logos, might count as a belief, except that it represents only a means of understanding reality. It may expose what things mean but not necessarily what should be valued. Rieff includes Freud among those who opted for theory instead of belief, going on to say that "the substitution of theory for belief is one of the highest achievements of secular scientific culture."[11] Yet can one subsist on theory alone? How and in what way does it nourish?

Rieff's candidate for the ultimate Freudian value is "honesty." Presented first by Freud as a therapeutic rule—enjoining full self-disclosure from a patient—its value lies in its ability to reveal needs the expression of which are otherwise censored by nature and culture. Therapeutically, honesty works to show the illusoriness of the gains of psychological illness; once exposed, the causes of the neuroses, hidden in secrecy, can be dealt with. "What is for Freud 'repression,' psychologically understood, is 'secrecy,' morally understood," Rieff comments. Secrecy is a category of moral illness, for it provides a hiding place for false motives."[12] Yet if an ethic of honesty promises a new ability to know the inner self, it also opens the way for a greater public knowledge of the private self. Freud felt that the result of such knowledge would be a greater toleration of the instinctual demands which would stand unmasked, but Rieff observes that it could also have the effect of making a person more vulnerable to the incursions of society. Nonetheless, individual freedom will in the long run make a gain, for it becomes possible for a civilization to arise which is not built on reticence or hypocrisy.

Rieff notes that "The ethic of honesty arose early in the middle-class era, and has been characterized by the modesty of its demands."[13] Yet he can go on to say that "Psychoanalysis has put into fresh expression the hard ethic of honesty, long in process of displacing the religious ethic. . . ."[14] But are the demands of honesty modest or hard? Rieff does not clarify the point and an implicit contradiction remains. It may be that they could be both. They are modest in that the emphasis of morality shifts from the adherence to rigid norms of conduct to an exposure of the instinctual life; a confession of instinctual demand absolves one from a moral obligation to certify as valid the demands made by society. Whether one can further succeed in practice in setting those demands aside is a separate problem; much will depend upon one's bargaining skill. But the ethic of honesty can also be hard insofar as the individual, having previously internalized the demands of society, will have to fight against himself, particularly against his sense of guilt, in order to bring his instinctual aspirations out into the open. On the basis of Freud's theory that the super-ego is a far more demanding master than society, the hardness of honesty would then consist in having the courage to overcome its demands; dethroning the super-ego can be a harder task than dethroning tyrannical kings.

A point of some importance, though, is that the individual will not be in a position to make his own ethic of honesty until he possesses what Rieff describes as a "pathology of moral aspiration."[15] One must, that is, come to see that the moral aspirations which weigh upon the individual, increasing his guilt, are derivative, being formed in reaction to instinctual desire. Once that is seen, and the essential sickness inherent in moral aspiration is uncovered, the individual can set about nullifying the internal pressures which these aspirations have generated. An understanding of the origin of moral aspiration becomes, therefore, a prime antidote to the pernicious power of those aspirations. "Freud," as Rieff puts it, "can conceive of a person's feeling guilty not because he has been bad but because, as a result of his repressions, he is too moral. This is one source of his influence: his diagnosis that we are sick from our ideals and that the one practicable remedy lies in an infusion from below."[16]

Rieff is clearly of two minds about the ethic of honesty. That it may help to end the reign of hypocrisy and deceit is to its credit, but it also leaves the way open to nihilism: "As a purely explanatory and scientific ideal, honesty has no content. Though the Freudian training involves intellectual judgment (it is, after all, psycho*analysis*), based on a calm and neutral appraisal of all the elements of a life-situation, still, the freedom to choose must end in choice. Here, at the critical moment, the Freudian ethic of honesty ceases to be helpful."[17] Moreover, there is no guarantee, Rieff believes—and, one might add, no logical connection—that acts which exemplify honesty with oneself will be good acts: "One need not be self-deceived in order to act maliciously. Freud gives us no reason why unblinking honesty with oneself should inhibit unblinking evil. Lucidity may render us exquisitely articulate and unapologetic about our aggressions."[18]

That Freud was himself aware not so much of the moral hazards here but of the psychological hazards—for not everyone could pay the price of honesty, which for some might increase their illness—helps take the sting out of the demands implicit in the norm of honesty; some compromise might be allowable. In the end, though, taken at face value, "The ethic of honesty does no more than establish the capacity to break the moral habits into which decisions, once made, tend to form themselves. Freud's is a penultimate ethic tooled to the criticism of ultimates. . . . Therefore, strategies of unbelief, of disclosure, of negation are themselves positive—effecting a studied release of energies heretofore inhibited by the ascetic character of the Western moral system."[19]

Yet it would be a mistake to believe that Freud was an uncritical champion of released instincts. He was too assiduous in pointing out the cultural need for repression to be cast in that simple a mold. His utopian ideal, when he had the individual only in sight, was that of instinctual freedom. But when he looked at the instincts themselves, he saw their intrinsic limitations; and when he looked at civilization he saw the importance of instinctual repression. If it is true to say that Freud was an antiprophet, holding out little hope, he drove his antiprophecy home by his systematic, theory-induced ambivalance even toward his own lack of a message; the best wounds, it seems, are those which are well salted.

It is possible to sense in Rieff's own ambivalence toward the ethic of honesty the stirrings of a competitive value system. When Rieff asks, "But what guarantees that an authentic action, or one conceived in perfect lucidity, will be good or in conformity with human nature?" he moves at once beyond the framework of the Freudian ethic he is interpreting. To ask what is "good" is to ask the kind of question which cannot readily, if at all, be encompassed within an intellectual system that would remain at the level of psychology or sociology. But this opening is not expanded. To have followed through here would have required that Rieff put very different questions to Freud. When Freud is presented as one who would have us reduce our ethical aspirations, renouncing our utopian hopes in the process, a countervoice is needed to explore other possible reasons why these aspirations exist. Freud may have taught us, as Rieff says, to renounce our highest hopes, to resign ourselves to the limits of our own nature and those of civilization, and to scale down our demands, but what if one wants more than this and is willing to pay the price for it? The Freud of Philip Rieff is detached, resigned, worldly wise. But cannot one decide to rebel against one's nature, choosing the way of risk rather than that of calculation? Rieff supplies no answer to these implicit questions, nor are they even raised.

Freud, he says, "is the architect of a great revolt against pleasure, not for it. He wrote no briefs for the pleasure principle. Rather he exhibited its futility. It is toward the reality principle that Freud turns us, toward the sober business of living and with no nonsense about its ease or goodness."[20] To this he adds, in a further passage, that "The reality principle does not completely supplant the pleasure principle, but it tries, through compromise, to exert discipline enough to meet the basic need for efficiency."[21] And then, after that, he says: "The practiced ease of expected disappointment recommended by Epictetus—beginning with a broken cup, and so on ultimately to one's broken life—is perhaps the best and most classical intimation of Freud's own way of coming to terms with life. To detach the individual from the most powerful lures of life, while teaching him how to pursue others less powerful and less damaging to the pursuer—these aims appear high enough in an age rightly suspicious of salvations. Freud has the tired wisdom of a universal healer for whom no disease can be wholly cured."[22]

In the last chapter of the book, "The Emergence of Psychological Man," Rieff introduces the conceptual protagonist of an antidrama which is to be fully staged in *The Triumph of the Therapeutic.* "In default of other cures," Rieff tells us in the opening page of that chapter, "egotism suits the age. . . ."[23] It soon becomes clear that, when Rieff speaks on his own, this egotism exerts a powerful attraction, though one essentially destructive in its thrust. The private man, whom Rieff saw Freud defending, is a man who both attracts and repels. "What is needed," Rieff states, "is to free men from their sick communities. To emancipate man's 'I' from the communal 'we' is 'spiritual guidance' in the best sense Freud could give to the words."[24] It is this "spiritual guidance" which Rieff carries through on, and he does it with conscious irony, in that way most approved by the social sciences: the word "ought" does not appear, only a set of analyses and predictions, the net effect of which is to stipulate the available choices and thus to determine the acceptable outcome. That much of this is, however, ironic on Rieff's part is by no means easy to detect.

"The successful patient," Rieff interprets Freud as saying, "has learned to withdraw from the painful tension of assent and dissent in his relation to society by relating himself more affirmatively to his depths. His newly acquired health entails a self-concern that takes precedence over social concern and encourages an attitude of ironic insight on the part of the self toward all that is not self."[25] The ultimate aim here is liberation, a liberation not from the self but from others. The ideal of community, with its built-in sacrifices, suppressions and suffocating moral ideals, must be overcome. If it is impossible to escape community, it is not necessarily impossible to fight a successful guerrilla warfare against it. The tools of this warfare are "honesty," the devaluation of ideals, the development of a tough, survival-conscious inner self, and the cultivation of a sense of irony to curb any tendency to take the outer world at face value. "When the inner life is not easily disturbed it has achieved what is to Freud as nearly ideal a condition as he can imagine."[26]

"Psychological man," the virtuoso of this inner life, is becoming the modal individual of our day. If the "derangement of communal life" cannot be overcome by social reform or manipulation, then the individual must look to himself for salvation. Rieff sees in this triumph of psychological ideals the presence of a tyranny:

> What has caused this tyranny of psychology, legitimating self-concern as the highest science? In part, no doubt, it is the individual's failure to find anything else to affirm except the self. Having lost faith in the world, knowing himself too well to treat himself as an object of faith, modern man cannot be self-confident; this, in a negative way, justifies his science of self-concern. Though the world is indifferent to him, the lonely ego may here and there win something from it. For the rectitude and energetic naïveté of the man who was the ideal type during the middle-class, Protestant phase of American culture, we have substituted the character traits of husbanded energy and finessed self-consciousness. The Frank Merriwell of psychological culture will not, like the moral athlete of

> Protestant culture, turn his reveries into realities. Rather, he will be mindful to keep realities from turning into reveries.[27]

The strength of "psychological man" lies not in ideals of right or might, or in religious or political creeds. He lives by the power of analytic insight which, if it gives him nothing else, supplies him at least with a strength to master his own personality. Freudianism, Rieff would have us believe, marks the end of man's battle with his own spirit. He knows how to turn reason to his own ends, to live beyond the requirements of conscience, and to approach with a wise skepticism the heavens sketched by politicians and economists. A new way of salvation has been found, one which systematically builds upon a rejection of all traditional modes of salvation. Once it is seen that there is no salvation, the individual is ready to go forth and seek what he can get.

In *The Triumph of the Therapeutic* Rieff carries through on these themes, only now increasing his own ironic style to a point where his hostility to the emergent "psychological man" is hardly concealed at all. The character norms delineated by Freud have now, for the first time in history, found a culture which can support them. The title of the book is itself a clue: the therapeutic approach to life, substituting analysis for ideals, theory for belief, detachment for commitment, and coolness for ardor has triumphed; there remains only the fulfillment of the triumph.

George Lichtheim has written of the triumph which Rieff details:

> The dissociation of personal from communal values is surely a basic feature of the kind of society in which we live. One may also call it the liberation of the individual from constraints not specifically laid down by law. This

> is an arrangement which liberals regard as an important achievement of Western society, and even Communists (at any rate in some parts of Eastern Europe) are now beginning to treat as desirable. . . . To grasp what is at stake it is necessary to stand back a little from the surface configuration of modern society and ask oneself what exactly it is that its citizens take for granted. It will then appear that what they take for granted is something very good indeed: namely, that there ought to be some kind of social morality—whether of a religious or a secular kind—but that it should not interfere with the personal freedom of the individual. For the citizen of modern society conceives of himself as an autonomous person owing no obligations to others or to the community, save for duties that are either contractual or specifically laid down by law. . . . It is for myself to decide how far I am prepared to exert myself on behalf of my neighbor beyond the constraint inherent in the prohibition to set fire to his house. In principle his problems do not concern me. What I do about them is "my own affair," and in no way dependent upon communal standards or shared values.[28]

That the "social morality" to which Lichtheim refers could now turn out to be survival is, as my last chapter tried to show, less paradoxical than one might think.

The key to Rieff's approach in *The Triumph of the Therapeutic* can be found in his understanding of culture. A culture provides two things. First, it provides a system of symbols whereby the moral demands men make upon each other for the sake of civilized life are codified. This codification, however, is only partially contained in legal codes and constitutions. The most important part lies in the unconscious, in the set of internalized norms and values which members of the culture imbibe from the overall atmosphere of the culture. Even the most ignorant will be bound to a culture, however little they may be aware of its formal structures. "Superior to and encompassing the different modes in which it appears, a culture must communicate ideals, setting as internalities those distinctions between right actions and wrong that unite men and permit them the fundamental pleasure of agreement. Culture is another name for a design of motives directing the self outward, toward those communal purposes in which alone the self can be realized and satisfied."[29] Cultures are marked by permissions and restraints, a Yes and a No. The relationship between this Yes and No is at all times dialectical. The present era of social change is marked by a shift in the balance of the dialectic. The shift is now toward the Yes, which means a shift toward the self, toward an acceptance of the rage against inherited moralities, toward a new way of organizing personality, a way more favorable to the individual: "The best spirits of the twentieth century have thus expressed their conviction that the original innocence, which to earlier periods was a sinful conceit, the new center, which can be held even as communities disintegrate, is the self. . . . Here literature and sociology converge; for the ultimate interest of sociology, like that of psychiatry when it is not lost in a particular patient, turns on the question whether our culture can be so reconstructed that faith—some compelling symbolic of self-integrating communal purpose—need no longer superintend the organization of personality."[30]

The predominant note in Rieff's description of culture follows Freud's: it is a system of remissions and releases, controls and restraints. Though it has what Rieff calls its "releasing devices"—means by which even in a traditional culture the individual is allowed some degree of tacitly permitted evasion of the demands of civilization—the essence of civilization is renunciation, control and sublimation. Freud took this to be the inevitable makeup of culture. Rieff, however, proposes that contemporary civilization believes it has found a way out of the Freudian trap. An attempt is now underway to create a culture which does not depend upon faith of any kind, nor upon the sharing of communal purposes. It is an attempt to fashion a culture where the shared interest in self-interest provides the only tie which binds people together. What Freud could never grant, that the individual could fashion a place for himself over against culture, Rieff argues has now been elevated to a communal goal—a society in which the shared value, to use a phrase which became popular in the late '60s, was that each person should "do his own thing."

The aim is a "full transition to a post-communal culture. . . . Now, contradicting all faiths, a culture of the indifferent is being attempted. . . ."[31] The means to attain this end are many, but each in its way contributes to the liberation of the individual from the restraints of culture: one aspires to think without giving assent, to know and not be deceived, to redefine good and evil and the nature of human perfection so that evil can be nullified by semantic decision. The old "corporate ideals" are not likely to see a rebirth: "By this time men have gone too far, beyond the old deception of good and evil, to specialize at last, wittingly, in techniques that are to be called . . . 'therapeutic,' with nothing at stake beyond a manipulable sense of well-being. This is the unreligion of the age, and its master science. . . . The systematic hunting down of all settled convictions represents the anti-cultural predicate upon which modern personality is being reorganized."[32]

Should we be tempted to think, hearing the talk of "community" and "commitment" in the air, that Rieff is wrong in discerning the signs of the time, he is ready with a response. For even "commitment" and "communal purpose" can be turned to the service of the self. The ingenuity of modern man lies in his clever use of faiths, commitments and commu nities—he denies their binding power by playing with them: "our cultural revolution does not aim, like its predecessors, at victory for some rival commitment, but rather at a way of using all commitments, which amounts to loyalty toward none. By psychologizing about themselves interminably, Western men are learning to use their internality against the primacy of any particular organization of personality. . . . Faith can then grow respectable again, as one entertainable and personal experience among others, to enhance the interest of living freed from communal purpose."[33] If "religious man was born to be saved, psychological man is born to be pleased."[34]

The symbol of the old culture was control, that of the new, release. The therapy of the old culture consisted in reconciling men to their culture, especially to its moral demand system. The therapy of the new culture will consist in freeing them, inwardly at least, from these demands. The tool of the new therapy is the "analytic attitude," which "expresses a trained capacity for entertaining tentative opinions about the inner dictates of conscience, reserving the right even to disobey the law insofar as it originates outside the individual, in the name of a gospel of freer impulse."[35] "Psychological man" is one who uses the analytic attitude in his own service; it enables him to withstand the culture, to remove himself inwardly from it, to defend himself against its incursions. By its anticreedal stance, the analytic attitude enables us to live without meaning and purpose. The great mysteries of life, which have befuddled the metaphysician and sustained the theologian, vanish: "Throw away all the old keys to the great riddle of life; depth in psychology brings men's minds around from such simplicities to the complexity of everyday tournaments with existence, to an active resignation in matters as they are, to a modest hope, and to satisfiable desires. . . . Psychological man may be going nowhere, but he aims to achieve a certain speed and certainty in going."[36]

The power of traditional culture—and all cultures until

the present were "traditional"—lay in its promise to man to secure for him a personal salvation and fulfillment on the condition that the individual identify himself with and subordinate himself to the community. It was a culture which required as its price of admission (and all had to pay) a considerable degree of asceticism, conformity to the common theology and public philosophy; together these added up to suppression of the individual. Such a culture placed a great emphasis on commitment to its ideals, faith in its ultimate purposes and acceptance of its personal demands. By the controlling mechanisms of the law, of psychological internalization and the presence of a social symbol whose contents were ascetic, self-denying and regulatory, commitment cultures were constituted. Yet they were, precisely because of their communal goals, values and shared public philosophies, positive in their thrust: they provided a matrix of meaning and hope. They offered not only collective security but also, for the individual, a warming communal identity. Well-being was constituted by membership in the community.

Precisely this kind of culture, Rieff argues, is coming to an end, to be replaced by a release culture, made possible economically by the spread of affluence and therapeutically by psychoanalysis. Well-being will now consist in a capacity to resist, not join, the community. Affluence makes possible for the masses the freedom from controls which was once possible only for the rich: "The faith of the rich has always been in themselves. Rendered democratic this religion proposes that every man become his own eleemosynary institution. Here is a redefinition of charity from which the inherited faith of Christianity may never recover. Out of this redefinition, Western culture is changing already into a symbol system unprecedented in its plasticity and absorptive capacity. Nothing much can oppose it really, and it welcomes all criticism, for, in a sense, it stands for nothing."[37] Scarcity made social cooperation necessary in the old culture; affluence will make it unnecessary in the new one. The technique of psychoanalysis makes possible the sustaining mentality by teaching people how to analyze away their ultimate concerns, to do without sustaining hopes and to cultivate that inwardness necessary to protect them against an internalization of the demands of the culture.

Together, affluence and the pedagogical techniques of psychoanalysis will make possible the sundering of the individual from the constricting demands of culture. Under the classical ideal, the healthy man was conceived to be the good citizen; private health was one with communal identity—"commitment therapies," whether secular or religious, sought to unite the individual and the community. Under the new ideal, brought about by "release therapies," the individual learns how to keep his identity separate from the community. A mutual conspiracy among all members of the society to give its blessing to this separation will make complete the transition, now in its remissive phase, from the old control to the new release culture. That there will probably always be some tension between controls and releases in society, and that the present cultural revolution is hardly complete may be admitted. But the revolution is under way, marked by the fact that the releasing symbol has grown more compelling than the older controlling one.[38] That the good life in our society is so often talked of in the language of "self-realization" amply confirms Rieff's point. It is a culture where the ideal is to be : "creative, spontaneous, additive, living in the now. . . . honest, truthful, individually autonomous, spontaneous, free."[39]

*The Triumph of the Therapeutic* ends in a prophetic outburst, in language which might suggest Rieff's approbation: "The therapy of all therapies is not to attach oneself exclusively to any particular therapy, so that no illusion may survive of some end beyond an intensely private sense of well-being to be generated in the living of life itself. That a sense of well-being has become the end, rather than a by-product of striving after some superior communal end, announces a fundamental change of focus in the entire cast of our culture—toward a human condition about which there will be nothing further to say in terms of the old style of despair and hope."[40]

But quite apart from whether the new release culture should be promoted, or even tolerated, is it realistic to expect its arrival? Can one be "neither hopeful nor hopeless but only realistic"? On Rieff's side are some obvious phenomena of contemporary life. Ours is a society geared to technological consumption. The basis of our economy is not only that people spend for the necessities but that they spend both for real luxuries and for those luxuries which a consuming economy has turned into social or psychological necessities. The success of a capitalistic economy requires a release of resources and a release of those acquisitive inhibitions which would compel a more security-minded people to save, store and accumulate. With so much psychological capital invested in status, a status now measured by the quantity of accumulations—cheap goods for the lower classes, culture and education for the higher—the psychological economy can only renew itself by ever more consumption, thus plowing the profits back into the consuming enterprise itself. All the while, technology proliferates, the primary vehicle for making consumption possible. Popular psychology, which has inherited only part of the Freudian message—the hazards of inhibition—promotes a life of release. The goods offered by "sensitivity training," the release of emotions, the enjoyment and acceptance of one's body, authenticity, self-fulfillment, all carry on, among a more affluent middle class, the work of release once available only to the more sophisticated and established classes. Buy—enjoy—live—fulfill—relax—swing: that is a litany, intoned in advertisements and the literature of psychological self-help, confirming Rieff's theory.

Even the immediate counterevidence is not as unambiguous as many make out. The regnant myth about contemporary youth is that of a new drive for community, a repudiation of the bourgeois prudence, technological obsessiveness and acquisitive drive of the old generation, a desire to seek an identity wider than that of the uprooted self, a vaulting idealism which puts the niggardly goals and self-calculations of the parental generation to shame. One might try to use this myth against Rieff, for it does imply that something new is happening in the culture and that, contrary to him, it is not psychological man who is being born, but who is dying. Yet however much white middle-class youth and their commercializing adult camp-followers enjoy cultivating this myth, it is not shared by those blacks who have been visited by the new idealists. The white kids who went south did so for a few summers only; the blacks who lived there saw this and could draw only one conclusion—the whites were doing it for their own self-fulfillment; not for them the asceticism, the commitment culture of one who comes and stays.* The New Left,

* Jonathan Kozol has made this point with great power: "I wonder, though, what kind of loyalties can be constructed on a groundwork of desertion. There is a black child I know in Boston who has now gone through four generations of white teachers, organizers, drifters, VISTA's, O.E.O.-supported revolutionaries and what he calls the hippie people, all in the course of six years. Peter can list the names of all the young white men and young white women he has known. They give him supper and they buy him shoes and take him on hikes and sit down on the floor and play with him for one summer and one winter, and sometimes for one spring and one summer once again. Then they switch gears, and they are into a New Thing. They cancel him out, or rather they do not 'cancel' him—they cannot quite do that—but situate him rather in a slot of history or a place of pain known as the 'race and conscience bag.' They make new friends and read new books and find a whole new set of words, and are off to a new dedication. . . . The instinct for this kind of nonstop forward

however much it has spoken of revolution, has at the same time resisted that discipline and authoritarianism which has been the historical sign of revolutionary seriousness. Only the blacks, desperate beyond the needs of their own psyche, have shown a talent for submerging their own identity and personal self-fulfillment in the service of their people. For the whites, any price will be paid for revolution but one—the loss of personal freedom of choice. Instead, the goal has been a mixture of communal radicalism and personal freedom.

The Hippies were a case in point of the way in which a movement that on the surface turned Rieff's psychological man on his head contained many characteristics of that man. While "community" loomed large in Hippie ideology—a sharing of possessions and women; a repudiation of dog-eat-dog individualism and competitiveness; communal living—individualism has been an equally strong note. Jesse Pitts wrote perceptively on the Hippies, and his description places them exceedingly close to psychological man: "For the Hippie all men are good if only they give way to their nature, as against the pressures of the social structure. Hence, uniqueness will be found in the cultivation of sincerity and spontaneity. Since all individuals are equally valuable and worthy of love, the desires of each merit realization. . . . One must always do what one wants to do; anything else is subordination and loss of independence, *ergo* loss of individuality."[41] This message, now well huckstered, has been taken up by the educated masses; see *Penthouse* and *Playboy*.

Even the concept of "love," affirmed by the Hippies but having no place in the canon of psychological man, had an ambiguity in Hippie communities which was not part of the

locomotion is present in all men and almost all women I know. . . . It is very, very attractive to keep moving onward. It is extremely hard to stay in one place and follow through on one thing." ("Moving to Nowhere," *New York Times*, March 24, 1972, p. 41.)

traditional concept. Missing is the traditional note of obligation, which entailed that love enjoined binding duties on each individual. For the Hippie, the obligation to love was universal, embracing all men. But so far as loving individuals was concerned, the requirement of obligation was sharply, fatally reduced by the countervailing obligation to retain one's independence. "A man and a woman," Pitts writes, "should be financially and emotionally independent from any other man or woman. This does not preclude developing attachments to particular individuals, but those attachments must involve complete reciprocity and never act as binding obligations upon the other. In order to be valid, a relationship must be totally spontaneous and sincere (there is a tendency to see these two kinds of behavior as synonymous)."[42]

In *Civilization and Its Discontents,* Freud complained that the Christian idea of love made unbearable demands upon people. The Hippie version, which retains the language of love, manages, by a severing of love and obligation, to scale down those demands: escape is always possible, structured into human relationships from the outset. No pressure of a relationship is so strong that it cannot be reduced or terminated by flight, now labeled as liberation and authenticity. And in other ways as well, the Hippie community blessed a movement from a commitment to a release ideology—in its emphasis upon retaining one's "cool," which amounts to preserving the autonomy of one's inner life; in its stress upon modes of release, whether through drugs or sex; and in its repudiation of work and occupational commitments with all the asceticism they carry with them. All commitments, in the end, can be seen in Rieffian terms as a playing with commitments for the sense of well-being that brings; but the lack of a permanent obligation to these commitments means that the possibility of guilt is reduced. Some of the sharpest criticisms of racism, war, urban poverty and the like have appeared in *Playboy*. Its implicit message has always been that radical social reform is perfectly compatible with the swinging life; there is a best of all possible worlds.

The question is not whether there exists a search for community, but whether the form of community sought is one which places the emphasis on corporate or individual identity. If the function of community is conceived as that of allowing the individual to find an identity in and through his relationships with others—seeing in "the other" a necessary source of an identity which transcends the private self—then one might say that the value of a genuinely corporate identity is being preserved. For the presumption of that kind of identity is that the isolated individual is incomplete; he will not be himself, nor will others be themselves, until a network of relationships is established. And that network entails binding obligations. That conception of community can be sharply distinguished from one in which persons join one another for functional self-releasing and self-realizing relationships only. The presumption there will be that people do need each other, but more for the sake of satisfying private ego needs than for personal fulfillment or human completion. The psychological-man model of community is the latter, with affluence diminishing the need for the kind of external assistance which dominated and determined the repressive contents of control cultures. Aided by affluence and psychoanalytical techniques, the enforced control community of earlier eras can be nullified. Thus will the individual be able to realize what he has always instinctually wanted—total autonomy and self-direction. Communal life becomes parallel play, with on occasion some rules being laid down to ensure that the players do not get in each other's way.

In the struggle between Eros and Thanatos, Freud saw the drive of Eros leading to an ever-widening scope of relationships; Eros seeks greater unity and human bonding. Biologically, then, the impetus of the life principle is not entirely in the direction of an individualistic satisfaction of instinctual urges—so much at any rate Freud argued in *Beyond the Pleasure Principle* and in *Civilization and Its Discontents.* The movement toward greater unity, entailing progress forward in the face of the backward-pulling power of Thanatos and toward a cumulatively greater range of relationships, is envisioned as a life principle because it nullifies the essence of the death principle—which destroys relationships and seeks a return to an original state of isolation and stasis. Rieff sees no evidence that any such power is now at work. "We are," he writes, "privileged to be participant observers of another great experiment by Western humanity upon itself: an attempt to build upon the obsolescence of both love and hatred as organizing modes of personality."[43] There indeed is tyranny, dressed up in a fresh new gown, but proclaiming as ever the obsolescence of what we have always known.

The irony with which Rieff analyzes psychological man makes evident his distrust and final rejection. But Rieff offers little to put in its place, in great part because he does not offer a positive view of culture which would strike a good bargain between the demands of the individual and of the culture. No more than Freud can he offer the foundation for a social ethic which would integrate a range of values in a way that would enable the individual and civilization to mutually behave toward each other in ways which respected the requirements of each. What Rieff has done is to lay bare the *hubris* and folly of an individualism run amuck, seeking a final break from all cultural restraints. But having rejected that form of individualism, what are the alternatives? Not an ethic of survival, which would manage to keep the individual in line at the price of a final victory of the community over the individual, resolving all tensions, ending the possibility of a mutual respect. If the tyranny of individualism, inherent in the mode of life of psychological man, presents only the prospect of a culture of self-contained human monads occasionally jostling each other, the tyranny of survival projects a world where the individual is effaced altogether. Both tyrannies are proof against any kind of social ethic, for both dissolve that necessary dialectic between individual and community which is the prime requirement of such an ethic. A failure in the first place to posit the validity of both individual and community will make it impossible in the end to combat the virulence of individualism and survivalism, a virulence which not paradoxically draws them closer together with every advance in technology and affluence.

The first step, then, in constructing a social ethic for technological societies is to reject the polarities of the analytic attitude, on the one hand, and the species attitude, on the other. The analytic attitude dissolves all of life into a cunning detachment of individual from community, providing the former with the psychological weapons to keep other human beings at bay. The species attitude, seeking only survival and perpetuation, provides no less effective weapons for keeping human beings at bay, only this time in the name of a future made safe for the future. The great threat to the possibility of a social ethic for a technological society is less the absence of all values than the triumph of one value over all others. Both individualism and survival are struggling to achieve that position, with a striking degree of success. Nothing is more important than to deny both the triumph they seek.

## NOTES

1. Philip Rieff, *The Triumph of the Therapeutic: Uses of Faith after Freud* (New York: Harper & Row, 1966), p. 261.
2. Philip Rieff, *Freud: The Mind of the Moralist* (New York: Doubleday Anchor Books, 1961), p. xix.

3. *Ibid.*, p. xxiv.
4. *Ibid.*, p. 3.
5. Robert Nisbet, *The Quest for Community* (New York: Oxford University Press, 1953), p. 7.
6. *Freud: The Mind of the Moralist, op. cit.*, p. 276.
7. *Ibid.*, p. 277.
8. *Ibid.*, p. 278.
9. *Ibid.*, p. 280.
10. *Ibid.*, p. 334.
11. *Ibid.*, p. 335.
12. *Ibid.*, p. 348.
13. *Ibid.*, p. 349.
14. *Ibid.*, p. 350.
15. *Ibid.*
16. *Ibid.*
17. *Ibid.*, p. 352
18. *Ibid.*, p. 353.
19. *Ibid.*, p. 354.
20. *Ibid.*, p. 355.
21. *Ibid.*, p. 357.
22. *Ibid.*, p. 359.
23. *Ibid.*, p. 361.
24. *Ibid.*, p. 262.
25. *Ibid.*, p. 362.
26. *Ibid.*, p. 376.
27. *Ibid.*, p. 391.
28. George Lichtheim, "Socialism and the Jews," *Dissent* (July–August 1968), pp. 333–34.
29. *The Triumph of the Therapeutic, op. cit.*, p. 4.
30. *Ibid.*, p. 5.
31. *Ibid.*, p. 12.
32. *Ibid.*, p. 13.
33. *Ibid.*, pp. 21–22.
34. *Ibid.*, pp. 24–25.
35. *Ibid.*, p. 31.
36. *Ibid.*, pp. 40–41.
37. *Ibid.*, p. 65.
38. *Ibid.*, p. 233.
39. Doris Grumbach, *Commonweal* (January 5, 1973), p. 306.

40. *The Triumph of the Therapeutic, op. cit.,* p. 261.
41. Jesse Pitts, "The Hippies as Contrameritocracy," *Dissent* (July–August 1969), p. 327.
42. *Ibid.,* p. 328.
43. *The Triumph of the Therapeutic, op. cit.,* p. 246.

## CHAPTER 6

# *The Control of Human Life:*

## *ETHICAL EXCURSIONS*

IT HAS BEEN FASHIONABLE in recent years to talk of a revolt against technology. The young, it is said, have turned to the counterculture for their new faith, substituting organic food, bicycles, and eastern mysticism for the established evangelism of science and its miracles. Even the older generation, if not quite into Con III, is less likely to take their technology without some hesitation than was the case with their parents. After all, the SST was defeated in Congress. There are two problems with this thesis. The first is that, if true, there could be a good possibility that some of the necessary and genuine progress which can still be attained through technology would be lost. Human beings continue to suffer untimely deaths from sickness and disease, and can still in many places use the advantages of better transportation, communications, sanitation and housing. Adaptation and preservation technologies, if not pushed too far, still have a key place in human life; it is difficult to imagine how things could be otherwise except at the price of great and unnecessary suffering.

The second problem with the thesis is that it is false. The great majority of young people have not returned to the land (which can hardly be found in any case), have not stopped making their living off a technological society, have not ceased buying automobiles, stereo sets, behavorial technologies, glossy magazines from high-speed presses, jet trips and instant telephone communications. On the contrary, there is every evidence that the symptoms of a rejection of technology are skin deep, one more cultural gesture of a generation (and its economic purveyors who know a hot item when they see it) which has not found, nor does it promise to find, an alternative. This is a generation which, while it has seen better than its parents the corruptions and excesses of the military-industrial complex, the tawdriness of a consumption ethic, the emptiness of chauvinism in the name of national power and prestige, has for all that been gulled most thoroughly and effectively into purchasing the most heedless anticommunal individualism, where the self and its liberation come before all else. In that respect, the sins of the father have been transmitted intact, with only slight changes in the decor. Its ready embrace of community is essentially anticommunal in its direction; community is for the good of individuals—one eye is always on the exit.

In a curious sense, the same inexorable logic which has been useful in struggling against war, poverty and racism—the perception that fundamental values were being paid lip-service only (justice, equality, freedom, for example)—has been no less useful in pressing toward the boundaries (if any) of personal liberation. And the ideological underpinnings of that liberation are none other than the same individualism which, less systematically applied, stands behind the whole edifice of modern technology. All that has happened recently is less a tribute to new insights into the hazards of technology by a more sensitive generation than a tribute to the power sustaining the very old myth of freedom from every cultural bond and shackle. But that was what the whole drive for more and better technology was all about in the first place; and a new resurgence of faith in technology can safely be predicted.

There is every sign that Rieff is right when he notes: "An old American alliance, between technological radicalism and cultural conservatism, appears broken. Battle-lines are being redrawn. The technological radicals will merge with the cultural radicals. 'Counter-Culture' is a purchasable gimmick."[1] Rieff does not present evidence to support his argument, but that is easily done. Perhaps the best evidence is the merging, on a mushrooming scale in the past few years, of counterculture themes with the latest techniques of the behavioral sciences. Increasingly, sensitivity training and human encounters have become both the object and subject of scientific manipulation. The magazine *Psychology Today,* part of a large publishing and media industry devoted to popularizing and commercializing the psychological and sociological sciences, beautifully illustrates the trend, mixing reports on the latest findings of behaviorism and humanistic psychology, surveys on the sex and other habits of its readers, interviews with scientific and cultural gurus, and heady reports on the latest trends at the self-realization front. That this package has been one of the publishing phenomena of recent years is hardly surprising, unless one has been naïve enough to think that science and technology are on their last legs. B. F. Skinner's *Beyond Freedom and Dignity* was a book generally panned by scientific and humanistic critics but one which, for all that, had a large sale and obviously met with more popular acceptance than all the talk of counterculture would have led one to expect (assuming one believed it in the first place). That *Psychology Today* gave over almost an entire issue to a lengthy excerpt from the book is no more surprising than the great amount of space it gave a few months later to the revelations of Don Juan, duly reported by Carlos Castenada.

This is only to say that there is no reason whatever to

think that a science of limits has become any less urgent because of counterculture outcries against technology. They are not serious outcries and, more than that, are likely to give way to a new burst of hope that technology will bail us out of our troubles. At the same time, if technology is to deliver the benefits of which it is capable, it becomes all the more important that it not be subverted either by excesses in its use or unrestrained optimism in its expectations. In Chapter 3, I argued that it will not be possible to understand either the possibilities or limits of technology unless there is some basic understanding of the human needs and desires from which it has sprung. That argument needs the complementarity of an understanding of the present cultural drama, which most prominently displays a convergence of fresh obsessions with survival and ever-heightened attempts to find a final epiphany of individualism, in which the self can operate without any cultural restraint at all. It is within this setting that a fresh reality principle must be worked out, one which is biologically, socially and psychologically sound; and on the basis of that reality principle a viable ethical response to technology found. Working with the themes of earlier chapters, I want in this chapter to lay out the rudiments of an argument for the social need for a normative ethic. The main requirements behind such an ethic are that it would be binding on consciousness and the cultural super-ego, able to cope with the twin tyrannies of survival and individualism, and capable of outlining a set of limits within which either to press forward or to hold back the development of technology as needed. That such an ethic will require a public morality to give it social substance, and thus to make a difference in the ways lives are to be lived and technology used, is intrinsic to the case I want to make. My focus in this and subsequent chapters will be on the technologies of biology and medicine, with which I am most familiar.

"Normative ethics," as we all know, disappeared some time ago, dying what was thought a richly deserved death at the hands of linguistic analysis, pluralism, and the "hard data" on the varieties of ethical behavior emanating from the social sciences. "Public morality," that foul invention of Inquisitors, Puritans, Comstockians and assorted thin-lipped religious dogmatists, was no less decisively put to death (at least in the better circles) by a combination of individualism, libertarianism, affluence and all forms of "liberation." I want to make a stab here at reviving the two corpses, or, to use my other image, to see if a bit of tolerance can be developed for those old and dirty expressions. I am not interested in seeing one ethical code developed which will, by law and social sanction, be *imposed* upon all. We know what that leads to, nothing good. Nor am I looking for, or would want to live in, a society which, even if not repressive, had reached that sublimely smug state of knowing *the truth* about good thinking and right behavior. At the same time, I think we are in less danger from that possibility at present than we are from the tyranny of individualism, which makes any kind of social ethic impossible.

Disclaimers of that sort, however, raise two questions. Is it even possible to think about "normative ethics" and "public morality" without, however unintentionally, taking a first step on the garden path? I can only say I hope so. More importantly, is there any hope that what one might genuinely want to call "progress" can be made, especially in an area which looks like a historical graveyard, full of dry bones? That is a question I cannot answer: we will have to see how it goes, and I can only say at this point that I think it would be foolish to predict an answer before an effort has been made. I would not even raise the topic but for the fact that the emerging ethical problems in technology, and especially in the life sciences, in medicine and biology, make it cry out for attention. A catalogue of these problems is not necessary, and only a few need mentioning. Increased longevity and highly sophisticated devices for the artificial maintenance of life brought to the fore a whole range of issues centering on the control of death, and thus the quality of life of those approaching death. Genetic counseling and improvements in the technique of amniocentesis, the imminence of mass screening for genetic defects, the prospect of *in vitro* fertilization, genetic surgery and cloning, all raise questions of individual genetic welfare and the future of human freedom, for political life, and for individual and public welfare. The need to control population growth, already touched on in earlier chapters, is still another example.

There are a number of distinctive notes about these problems, which taken together point to the need for a reexamination of normative ethics and public morality. First, a number of advances, primarily in genetics, could have a powerful impact on the lives of future generations, much more than on our own. The choices we make now may either increase the welfare of future generations, or seriously and perhaps irreversibly limit it. Two or three generations may be required to determine the consequences of genetic manipulation. Second, the increasing complexity of the medical and biological developments means that the public, for the most part, will be only dimly aware of their social meaning and implications. They will find it increasingly difficult to judge whether a particular scientific advance or technology is good or bad, both in the short- and in the long-run. Third, the possibility of different and discrepant ethical standards existing between the medical and technological community, on the one hand, and the larger community, on the other, threatens serious problems of mutual trust. There is already some evidence, for example, that a majority of physicians believes that most dying patients *do not* want to be told they are dying, while a majority of the public says that they *do* want to be told. In many cases, since patients are dimly aware of these different standards, the result is a distrust of the physician, whom they suspect may not be telling the truth. Fourth, and perhaps most subtly, much of the medical experimentation on human subjects necessary to develop future technologies will be carried out on a "voluntary" basis. But the social consequences of these experimentations, however much coercion of individual subjects may be absent, raises a whole range of serious problems. They cannot be ignored simply because some subject voluntarily signs a consent form.

Why do these problems raise issues of "normative ethics" and "public morality?" The first step in answering this question is to make some observations about the way many appear to make ethical decisions, particularly those in the scientific and medical community. My observations are based on a reading of scientific literature and on personal experience, not on hard survey data, which does not exist.

A number of schools of popular ethical thought can be distinguished. The first can be called the "religious" school. Here, ethical decisions are made out of the context of some tradition of religious morality. If asked for a reason why he thinks some particular act is "right" or "good," a person in this school is likely to reply, "because I am a Christian," or "because I am an Orthodox Jew," and so on. The second can be called the "emotive" school. Those in this school believe that certain acts are right or wrong because they "feel" they are. "Gut reactions" are taken to be normative, and, if reasons are requested, the response is simply to report again how one "feels." The third school can be called the "conventionalist" school. Acts are judged to be right or wrong on the basis of accepted conventions or mores. An acceptable reason for a certain type of behavior is, for instance, because "that's what everyone does," or "it's always been done that way." A fourth school might be called "empirical conventionalism." That school relies on the results of public-opinion surveys. If a majority of people think something is right, then it must be right. An acceptable reason for one's behavior is the latest survey showing that a majority agrees with you. A fifth school is that of the "simple utilitarians." Acts are good because they are believed conducive to the greatest good of the greatest number. An acceptable reason for this school is one which shows that more people will be helped than harmed.

If these are representative of the kinds of reasons given for personal behavior, another and related cluster of schools can be found for making judgments about public behavior and the resolution of social conflicts. One of these is the "barefoot civil liberties" school. Everyone should be free "to do his own thing" or "to make up his or her own mind." It is not necessary for people to have to justify their conduct; it is enough if they are "honest," "sincere" or "authentic." No one has to give a reason to anyone else for anything. Another school is that of "gross majoritarianism." Those public acts and laws are right which command majority support; or if it is legal, it must be morally acceptable. The statement that some act is "not against the law" is often put forward as an ethical statement in this context. Still another school is that of "primitive cost-benefit analysis." If it can be shown that the public will save "x" dollars by carrying out a certain policy, then that must be the ethically correct policy.

Finally, one should mention a variant on "professional ethics." I am not thinking here of special codes of conduct for particular professions, but rather of those popular reasons many technologists give to justify or excuse what they do. I will just cite some expressions here: "that's not our responsibility," "if we don't do it, then someone else will," "I'm not a philosopher or theologian," "that's a political (or ethical, or social, or theological) question."

I wish I could say these descriptions are nothing more than parodies. But every one is an expression I have heard repeatedly in discussions with scientists and physicians, and they abound in the scientific and medical literature. Moreover, the kinds of expressions I have quoted often represent the sum total of an "ethical position." No further explanation, or any fully articulated defense, is forthcoming. Let me say at once that I do not mean to be *entirely* condemnatory here—just heavily so. It is rare for even the most highly educated and sensitive person to be able to provide a full account, well defended, of his or her ethical positions. Even professional philosophers and theologians, who can often enough speak knowingly and critically in the abstract of the strengths and weaknesses of different schools of ethical thought, are not necessarily all that articulate or rational when it comes to explaining why they think some specific act which involves them personally is right or wrong.

My main point, then, is not to bemoan or excoriate the failure of most people, including scientists and physicians, to be able to present well-reasoned, plausible ethical positions. I only want to call attention to the great diversity of popular ethical opinion, and to the fact that, philosophically speaking, much of this ethical opinion is simply *held* and, apparently, acted upon without benefit of any felt need to provide a full rationale for holding the opinion. In a word, most people seem to make do with an ethical slogan or some kind of one-sentence general principle. In one sense, it is a tribute to the political freedom in our society that there can be such diversity and so little pressure on its citizens to justify their ethical decisions in a public and critical forum. But at the same time we need to ask whether this incredible welter of opinion, together with the absence of pressure to defend and justify, provides a healthy setting for coping with ethical issues of technology. The sometimes spoken and sometimes unspoken premise is that, ethically, one position is as good as another, a kind of easygoing ethical geniality which is politically highly serviceable in our kind of pluralistic society.

Indeed it is, up to a point. The question is, "What is that point?" It may be objected at once that even to suggest there is a point where the ethical diversity, informality and freedom cease to be valuable is to raise the specter of the dictator or the public imposition of "correct" thinking. But while history has ever taught us the hazards here, it is no less true that few of us are willing to really follow through in allowing everyone to act as he or she sees fit. We cannot do so. Most of us do not find freedom served by a "live-and-let-live" tolerance toward anti-Semites, racists, government manipulators, industrialists who pollute the water and air. We are *forced* into this position, if only because it is rarely possible to honor all values simultaneously. The demand of the racist that his freedom and opinions be respected and allowed full circulation is rejected—because we know the price racial minorities will have to pay in terms of their freedom, and the justice owed them, if the racist's claim for full freedom and equality is honored. We cannot allow industrialists full freedom to destroy forests and pollute rivers, nor individual seekers of release the liberty to rape, murder and steal.

Let us consider, in the context of the foregoing remarks, some of the more pressing issues posed by technology. How are we to determine, and then act upon, our obligations to future generations? Reflect, in particular, on the possibility of present genetic interventions which may significantly and irreversibly influence the genetic constitutions of future generations; nonetheless, they will have to live with the fruits, hopefully good, but possibly bad, of our choices. Is it sufficient in this kind of situation to let the matter be decided by individual scientists, those technically capable of making the intervention? What would we be willing to count as a good ethical reason for running unknown genetic risks with the welfare of future generations? That some people *feel* it would be acceptable? Or that by their private reckoning it would contribute to "the greatest good of the greatest number"? Or that, since the topic is controversial, each scientist should be allowed to make up his or her own mind? Or should the matter be put to public vote, going ahead with the intervention if a majority approves?

I am not interested at the moment in judging the validity of any of these possible responses. More fundamentally, I am concerned with whether the "ethical geniality" I mentioned above would constitute responsible behavior on our part as a community, in the light of our possible obligations to those as yet unborn. I seriously doubt it. At the very least, future generations would have grounds for condemnation of our behavior if they discovered that the matter had been decided on the basis of the unexamined "feelings" of a small minority. A solution which allowed each scientist to make up his own mind, willy-nilly affecting the lives of others who will have no choice themselves about the consequences, will seem no less capricious.

The question of different and discrepant ethical judgments between physicians—technological experts on the care of the human body—and patients poses no less vexing difficulties. Part of this problem can turn on different standards of what is good for the individual, with a conflict between the patient's judgment of his case and the judgment of his physician. Another part can stem from a conflict between a physician's judgment of what he owes society over against the claims of individual patients. It is increasingly common among genetic counselors, for instance, to worry whether their obligation to the genetic welfare of society and the species should take precedence over their traditional obligation to individual sufferers of genetic disease. Some patients might well believe themselves victimized if chance should drop them in the waiting room of a physician who gives the benefit of doubt to society, at their cost. The care of terminal patients provides another example. Some physicians are willing enough to cease useless treatment, to allow "death with dignity." Others, however, feel compelled to use every bit of their technology to the bitter end, despite the wishes of patients and their families. Is it proper that, once again, chance may dictate which kind of physician one gets? Or that, because of the power given the physician, the choice will be his rather than the patient's? And is it good, in the end, that so little social consensus exists on what is right and proper for everyone, physicians and patients?

The most difficult problem is the extent to which society ought to rely upon the norm of "voluntary consent" as an ethical justification for medical experimentation on human subjects. At one level that norm is a critical safeguard for potential experimental subjects. Their bodies may not be tampered with without their permission, no matter how worthy the scientific cause. Of course there are abuses, and the pervasive skepticism about the possibility of truly voluntary consent is sensible. Even so, the existence of such a norm is valuable for the protection of individual rights. At quite another level, however, one must ask whether every kind of experimentation to which consent might voluntarily be given is necessarily good for the individual giving the consent or for the society. Is it ethically good that some people should allow psychosurgery to be performed on them, at the price of irreversible personality changes, or that they should allow the implantation of electrodes which permit their behavior to be controlled by another person? Is it morally desirable that some women should be allowed to consent to contraceptive experimentation which may result in deformed children? I am citing very uncommon cases here, but each raises serious questions of the extent to which a pure ethic of free choice should be allowed to determine the outcome. We live in a society, unfortunately, where with enough diligence one can find someone who will consent to anything and everything. But do we want people voluntarily to consent to anything and everything? Various methods of biological warfare, we might recall, could not have been developed had there not been willing experimental subjects.

Naturally, each of the cases I have presented here has its other, and more beneficent, face. In the case of some genetic diseases, it is quite possible that the physician has a greater obligation to society than to the individual patient. It may be just as well that physicians are not always obliged to do what their patients ask. If an earlier generation of experimental subjects had not freely consented to some hazardous medical experimentations, many of us would not be alive today. Nothing I have said should be construed to suggest any clear-cut ethical resolution of the problems raised. The dilemma is more fundamental than the particular issues. How should we as a society cope with these problems, when it seems that hazards await us no matter where we turn? Most critically, can these problems be solved by means of a do-it-yourself ethic?

One commonly proposed solution is less adequate than it appears. That is what might roughly be called a political solution: increasingly, in an era which has grown at least wary of uncontrolled technology, it is proposed that there be public review and discussion of actual or potential scientific developments. "Technology assessment" has emerged as one response, together with various schemes to improve scientific "peer review" procedures by enlarging their scope to consider ethical and social implications. The general concern is that a way be found for the public to have a greater role in the establishment of science policy, in the determination of desirable lines of research, in the application of new technologies, and in the development of professional codes of conduct. The spirit behind this concern is commendable; the public should be consulted far more than it has been in the past. "Science for the people" is not a meaningless slogan.

Realism demands, however, a recognition of the limitations of the possibilities here. One limitation is the fact that considerable education and sophistication are often required to understand what is or may be at stake in biomedical and other technological developments. When even those in different branches of science cannot keep up with or understand what their colleagues in other specialties are doing, it is asking a lot to expect the ordinary citizen to do any better. I am not utterly pessimistic here, but at the least one should realize that a very formidable education program would be needed, and not once but again and again. Moreover, the reality of much scientific work at the frontiers is that its scientific status is often hotly debated among scientists themselves; try, for instance, to get any consensus among biologists on whether and when it will be possible to "clone" human beings (i.e., to produce an unlimited number of genetically identical persons).

Another limitation is of a different, but no less important, order. The progress of science in the modern era has been the fruit in great part of the freedom science has enjoyed, especially its freedom from ideological and governmental control. Wherever science has been subject to governmental sanctions it has suffered. Even if the notion of "the people" controlling science is far more attractive than that of totalitarian governments, the results might be equally deleterious for the quality of scientific work. Even now, basic research has been harmed in America by popular pressure in and on Congress to fund only "relevant" research. Still another limitation is that, unless there is careful thought given to the ethical issues before decisions are made by the public, there is no guarantee that the decisions will be wise.

That last point will serve to reveal my own hand. It is quite possible that, in handling the ethical problems of technology, the democratic political method is the best we have and all that we should aspire to. It does provide a procedure for resolving public disputes, and, together with the courts, ways of adjudicating conflicting values. If that is the case, then the best path to follow would be to attempt to maximize public information and debate, submit vexing issues to courts and legislatures when necessary, and hope for the best. Ultimately, I believe that is the wisest and most effective method we have for coping with different viewpoints in the public arena.

At the same time, if we are concerned not only with using the best available political methods, but also with trying to act toward the ethical problems themselves in a wise and responsible manner, then something additional is needed. As a starter, I do not believe that, in fields as intimate as biology and medicine, good laws and professional codes will suffice to produce sensitive ethical behavior. Everything cannot be legislated and, in any case, science and medicine might well be badly crippled if an attempt was made to bring all acts under legislative or legal control. Barring that kind of control, by what means should individuals and professional groups try to work through the ethical problems? What should their philosophical methodology be? What sources ought they to draw on for their thinking, and what criteria for judgment ought they to use? If, in many cases, some kind of ethical agreement will be necessary, even if only of a rough kind, on what should this ethical agreement be based?

Questions of this kind—which imply that good political and legal procedures alone do not suffice—ineluctably raise normative issues. Moreover if it is agreed that, although law and legislation cannot or should not be allowed to control all conduct, good behavior is nonetheless still desirable, how is such behavior to be facilitated and encouraged? And what should be counted as *good* behavior? That concern pushes us into the problem of public morality, by which I mean the standards which govern those private acts either of individuals or of groups which have social implications for the welfare of others. It is simply not enough when, as in medicine, one person has the power of life and death over another, to say that the only ethical requirement to be laid upon the physician is that he act according to his conscience, or that he should be allowed to make a free choice. It is his *acts,* not his intentions, which will cure or kill us; so far as possible we want his decisions to be *right,* and not just conscientious. Future generations will care less about the procedural impeccability of our present decisions which will affect them than they will about the quality of the decisions.

The very idea of a normative ethics entails at least two important assumptions. One is that society, particularly in confronting ethical issues of technology and the life sciences, would profit from an attempt to develop some common core of ethical standards by which to frame the problems, and to set some outside limits to acceptable answers. If that attempt succeeded, even in part, elements of a public morality might be found. The second assumption is that the effort would not be in vain. Concerning the first assumption, as I hope my earlier comments have shown, the attempt would be valuable. We cannot claim ethical seriousness or responsibility if we do no more than blithely and passively accept the ethical geniality I pointed out above. Concerning the second assumption, I see no reason for complete despair. There are certain values most people share, even if they do not always agree on their meaning or application. We do know more than earlier generations about what people need, at both the physical and the psychological levels. There still exist many traditions which, if in need of some refurbishing, are by no means dead. We have had in recent centuries some very bad experiences, which might tell us a thing or two. There are, in a word, some starting points. Where they will take us we will not know until they have been followed through.

In *The Triumph of the Therapeutic,* Philip Rieff argues that "a culture survives principally . . . by the power of its institutions to bind and loose men in the conduct of their affairs with reasons which sink so deeply into the self that they become commonly and implicitly understood."[2] In this vein, he goes on to quote Harry Stack Sullivan's observation that the dynamics of a culture are in "the unwitting part of it."[3] I quote these passages out of a conviction that a satisfactory resolution of the troubled relationship between ethics and the life sciences, and technology more broadly, must be cultural, not individual. The culture which will find a valid resolution does not yet exist. Let me define that new culture as one in which human beings can live (and die) securely, harmoniously and humanely in the presence of constant technological advances in the medical and biological sciences. While this new culture will no doubt have its articulated norms, codes and regulations, the test of its worth and of its staying power will lie in "the unwitting part of it," where most people actually make their decisions and live out their lives. This is not to deny the importance of "reasons" which lie on the surface, the stock-in-trade of moralists, legislators and pundits. But it is to deny that they will be sufficient.

They will not be sufficient, on the one hand, because it is exceedingly difficult to articulate moral rules and legislative norms which can take account of the complexities of concrete decision-making. It is not difficult to note the inadequacies of such general principles as "the greatest good of the greatest number," or "do no harm," or "do good and avoid evil," even if, at the same time, ethical thinking can not make a beginning or an end without such principles. They will not be sufficient, on the other hand, because it is doubtful that most people act in terms wholly of well-defined moral codes. Man is more than just a rational animal; he feels, senses, imagines and relates. A culture requires for its sustenance the presence of "reasons" which have sunk so deeply into the self that they inform the a-rational side of man as much as the rational. That is the whole point of Freud's cultural super-ego.

The "unwitting part" of the moral life lies in the spontaneous precritical responses we feel, in what repels and attracts us. Reason can provide a check and a corrective to these responses, but it is not our sole guide to the moral life. It is no accident (nor necessarily a sign of a decadent culture) that if you ask someone what he "thinks," he is more likely to tell you what he "feels" or "believes." The more serious and deep the question is, the more likely that response is. Why should it be otherwise, if man is more than just a mind? (That kind of response does not of course serve as an automatic justification for beliefs or behavior, which is another matter altogether.) I am not saying that all moral judgments and ethical codes can be reduced to feeling and instinct, with reason no more than an instrument of rationalization. That view has always required a faith and a model of man which flies in the face of experience. My point bears on the nature of a viable culture which, in its images, visions, institutions, rituals, biases, and unwitting responses reflects the full range of human capacities; and, in doing so, allows people to live a meaningful life. The ethical part of that life will partake of the whole culture, not just of its formal codes. A culture is a nurturing context, feeding and shaping us.

It is often asserted that our culture neglects to ask the most basic questions: What is man? What is the good life? What is really harmful and what is really beneficial? The observation is surely correct, just as it is correct to say that a culture which does not, from time to time, take stock of itself by asking such questions is a diseased society. But there is also a sense in which these are unbearable questions, bringing no solace, yielding no ready answers. A healthy culture may be defined as one where they need not become an obsession. Who could blame the person who would shy away from them? Except for those who dote on such things, who could not fail to resent being pushed to the wall on the nature of ultimate truth as the price of daily living? Everything should not be demanded of everyone. One function of a culture is to make up for our deficiencies in private wisdom, to take some of the burden of truth-seeking off our individual shoulders, to save us from going to the mat on each and every question, to allow us to know things in our bones which we do not by philosophical standards know in our heads. To be sure, every individual will have to seek the truth at some time. But he should not have to do it alone and he should not always have to do it. A culture should assist him.

Unfortunately, we do not live in such a culture. We are thus forced to ask impossibly hard questions, and to sit around tables inquiring, "What is truth?" In the instance of medicine and biology, the problems are compounded. By and large, biomedical advances have relieved man's estate and added to his knowledge. Our culture has lived comfortably with the belief that the future would only bring more of the same—better health, longer life. The ethos sustaining this belief was comforting and coherent; people could content themselves with being good journeymen in carrying on the splendid fight, thrilling to each new scientific discovery and each new technological application. Ultimate questions did not have to be asked every day. They had once been asked and answered. What remained was the less anxiety-producing business of carrying out the details of the new gospel of unlimited technological progress, by now sunk deep into the unwitting part.

The new gospel has run into trouble, beset by a plethora of

undesired medical, ecological, behavioral and demographic results. Some now question its root and branch; their rejection is only thinly disguised, springing from a fundamental fear of man's dehumanization or self-destruction. Others, not sharing that apocalyptic vision, hold firm to the old faith, but a more sober faith, now tempered by an acute awareness that the new medical and biological powers can be misused or give forth some unintended destructive by-products. Still others are simply in a quandary, attracted by the possibility of remaking man but uneasily aware that such a project is unprecedented. None of these groupings might exist of course were there not others, steady in their faith, working away as diligently as ever to cure people of their bodily ills, to remove some flotsam and jetsam from the gene pool and, in their hopes and by their lights, to do a better job with man than an untutored and uncorrected nature has been able to do.

The net result of these cross-currents has been to jar loose from their subterranean moorings those "reasons" for going forward which had sunk out of sight. They are floating to the surface and being inspected. That event in itself signals the dissolution of a culture. The question is how great the dissolution will be, and the extent to which a new culture will take its place. I would guess that the breakup will be limited. Having tasted many of the fruits of a serene faith in unlimited medical and biological progress, humans are hardly likely to dispose of it altogether, however much the serenity may have to be modified. People are not likely to get over their inveterate distaste for sickness, disease and death. Nor, more profoundly, are they likely to adjust themselves happily to the limitations their finite bodies impose upon them.

The quest for transcendence is endemic to human nature. Cut away everything else, and that is the motive lying behind the scientific enterprise, to go beyond life as we find it. To take the most utopian vision, centering on the possibilities of genetic and behavioral engineering, what is it other than one more attempt to take arms against a sea of troubles? One would be less than human if no such visions entered his head.

Yet, no less clearly, if the idea of progress is not likely to be abandoned, it cannot go on in the way it has manifested itself in the past, wildly, recklessly, heedlessly throwing up a profusion of uncontrolled goods and evils. This reckless profusion comes precisely from the fact that the original reasons for wanting scientific progress have sunk too deeply into the self to allow the kind of periodic reexamination and correction they require. Progress was possible because attention and energy could be shifted from justifying the very idea of scientific and technological progress (which took some doing in the first place) to getting on with its implementation. For well over a century that path has been explored, and presumably we have learned something. Yet if a reexamination of deep premises and fresh inspection of high visions is in order, periods of that kind of acute self-consciousness are usually short-lived. One can't just talk forever; there must be some living and some acting. Eventually "reasons" will be shaped and they will sink below the surface, for another generation (or two or three) to live by and with in an unwitting way. That means also that something more is required than coming up with good "reasons." It also requires that a cultural base be established which will, in its shaping of people, instill in them good habits, sensitive emotional responses, good desires and healthy repugnances. The reasons will establish a direction and some general standards; but they are not the whole game.

To build a fresh ethic for technology and the life sciences is to build a culture. Cultures consist of people working for common ends, sharing some binding visions, agreeing on some common values. They are erected on the scaffolding of laws, rituals, customs and public institutions, which together give shape and strength, concreteness and specificity to the visions and the values. To be viable, a culture must provide a nurturing ground for the development and enrichment of human life. If either the values are corrupt (however fine the laws), or the laws and institutions are moribund (however solid the values) that development and enrichment will be threatened.

Most critically, a culture will have to offer ways of meeting some basic human needs, needs which go beyond physical well-being. Human beings need a sense of *meaning,* by which I mean a way (minimally) of coping with absurdity or (maximally) of developing a system of belief which makes sense of things. Few can live with Freud's harsh ethic of honesty (however easily they can live with its benignly deceptive contemporary forms). Questions always have and always will be asked about the meaning of life; they can be dismissed only temporarily. No culture as such can answer those questions, and one would be justified in a deep suspicion of any culture which claimed it has or could provide all the answers. But a culture should at least provide some of the answers, and a setting in which those as yet unanswered can be pursued. Inevitably the institutions and the mores of a culture will reflect some latent answers to the question of meaning; roots can be found, however hidden. Moral codes and ethical systems develop out of the matrix of theories and convictions concerning the nature of human life. Any attempt to develop or sustain moral codes without such a matrix will in the long run fail, either because apart from fickle custom there will be nothing to nourish them or because, unbidden, the codes will themselves have the tendency to create structures of meaning, which at some time will demand examination as their concrete implications unfold in experience. In some fashion, then, a culture must find a way of coming to grips with the drive for meaning, and it must find a way also of bringing about a coherent relationship between that drive and the particular moral and legal codes it devises to manage the relationship among the people who live in that culture.

Together with a sense of meaning, human beings also need a sense of *community*. By that I mean the possibility of living in peace and mutual enrichment with others. For community to exist, trust is fundamental, a trust which springs from shared values, an interdependent life in common, and a sustaining belief that in the ordinary course of events one's neighbor is not one's enemy. The physician-patient relationship, for example, is utterly dependent on trust. Observed laws, whose spirit and not just letter are accepted, provide one measure of community. At the same time, there must be standards of value, of right and wrong, which exist independently of the law. Without such standards, there would be no way of judging the human validity of law. Most importantly, mutual trust will be displayed by a feeling of security in the presence of others, particularly in the presence of strangers about whom we know nothing other than their common citizenship in our own society. To say we know what to expect from others in a genuine community takes us part of the way toward that feeling of security. Even more to the point, though, we need to know *why* the expectation of security and safety exists, an expectation which needs greater support than the mere existence of protective laws. That depends upon an expectation, supported by experience, that certain values are shared, even by those whom we do not know in any personal sense.

If the drive for meaning and community are strong, they are no less strong than the drive for *values*. By this I mean the need for coherently structured goals, which allow both individuals and communities to determine what they should seek and what they should avoid. Human desires, in the raw, are conflicting and contradictory. They have to be sorted out, and a distinction made between what at any given moment is in fact wanted or sought and what ought to be wanted or sought. One cannot for long live with himself unless some sort of internal order has been established, with priorities, ways of resolving dilemmas, and a reasonably settled sense of where the good of selfhood lies. Freud's "ruthless honesty" is finally inadequate to this kind of task. On the cultural level, life in common requires a basic compatibility between and among the values of different individuals. Beyond that, as suggested above, there must be some common social convictions of where the good of each and all lies.

Finally, there is the drive for *integrity*. By this I mean a drive for a unified self, one certain of its own worth, at ease with itself. Our culture of late has laid a heavy emphasis upon self-determination as a critical determinant of integrity. But it is not fully clear to me whether the individualism which this emphasis bespeaks stems from psychological richness or poverty. In its most extreme form, this individualism presupposes that, because there exists no ultimate basis for the grounding of a drive for meaning and values, each person must create his own world *de novo*. Thus self-determination becomes the basic good to be sought, as the essential condition for the creation of self-sustaining private worlds. Curiously, while laissez-faire capitalism is on the decline, laissez-faire self-creation continues to find an audience; psychological man is coming into his own. Both assume that there is some natural marketplace in which a hidden hand harmoniously regulates private and public interests. While that illusion has been debunked in economics, it hangs on and is currently flourishing in the realm of popular values.

I have stressed the drive for meaning, community, value and integrity in order to establish a ground for debate. Put plainly, my contention is that the development of an ethical system capable of managing the issues thrown up by developments in technology, and the medical and life sciences, must take account of and integrate each of these drives in some coherent overall perspective. That perspective must, in turn, be embodied in a culture, suffusing its laws, customs, institutional arrangements, political life and super-ego. It is clear that, at the moment, there exists no common perspective for dealing with the ethical issues of the life sciences, or with most other advanced technologies.

One example should make my point clear. The question of which sufferers of kidney disease should be chosen for renal dialysis when there is a scarcity of dialysis machines is, by now, one of the prototypical dilemmas used to illustrate the kinds of problems posed by advances in medical technology. My assertion is that this kind of problem cannot be answered culturally unless the solution embodies a coherent view of the meaning of life, expresses some acceptable understanding of the mutual obligations of life in common, displays an ordered, reasonable and deep commitment to some clear set of values and, finally, does not threaten either the integrity of those who are chosen or those who are not chosen for the machine. A failure to meet even one of these tests will result in a failure of any given answer to the question, "Who shall live and who shall die?" The failure will come because those who are the losers will feel they are victims of an absurd universe (once again, on top of their kidney disease), or of a discriminatory society, or of the power of those with values different from their own, or because their own integrity (which at least involves a will to live) has been violated. To be sure, I am specifying a heavy burden to be met by any proposed solution. But it is at least advisable to be clear about the ingredients of a good solution, which must, in any case, elicit a deep and positive response in the culture.

Both the public and the scientific community are now forced to get by on this and similar questions with a set of reflex responses, crudely adapted to the needs of the moment. One set of responses takes the form of pop-pragmatism, summed up by the ubiquitous statement that "I'm just a practical person." Translated, this usually means doing whatever convention (old or avant-garde), the zeitgeist, and the opinion of one's peers will allow one to do. The difference between the philosopher and the "pragmatist" (or "practical man") is that the former knows he is philosophizing while the latter has deceived himself into thinking he is not (both may be right or wrong in their philosophy—but that is a separate problem).

Another set of responses goes in quite the opposite direction. In the name of a supposedly more sober and thoughtful consideration, current practice and trends are rejected, accompanied by invocations of a "natural" standard, usually based on an inarticulated notion of the essence of the human. Too often it turns out, however, that the proposed "essence" amounts to little more than a resurgence of the outlook and conventions of some earlier generation; fresh and illuminating insights into the nature of man are not provided.

I have drawn two caricatures here, but not far removed from the reality. In a time of cultural transition, public debate usually takes familiar forms, with one side bemused and thrilled by the new, and the other hanging on tenaciously to the old, fearful of the decline and fall of everything sacred. We are caught right in the middle of such a debate, and the fate and future of our culture will be determined by the way the weight of opinion falls, by the kinds of reasons which are allowed to sink deep into the self. Since we are, whether we like it or not, engaged in the task of building a new culture, it is important to determine where the major issues lie. Only then can we begin thinking clearly about the proper "reasons" which should shape the new culture.

Three issues are of overriding importance. The first is the nature of man, what he is and what he can become. The second is that of the extent to which nature (human and nonhuman) can and should be manipulated and controlled. The third is the relationship between public and private morality, law and ethics. This is to say no more than that a culture requires some image of man to work with and to hang its hopes on, some image of man's relationship to that empirical reality which he finds to be his environment, and some image of the relationship of the personal self within to all those other external selves who are our fellow human beings. I use the word "image" to make clear my belief that purely rational, verbal formulations can never do justice to the full reality of any of these issues, either descriptively or normatively. Or, put another way, "reasons" should be understood as a complex mixture of thought, feeling and imagination.

It is specifically when one considers *the nature of man* that verbal formulations fail. I cast some doubt above on those who invoke a nonarticulated quintessential nature of man as a club with which to beat those who speak too easily of creating a new and better human nature. But I do so out of a certain despair. Nothing would seem to me more desirable than to be able to specify some fixed, normative human nature, by which one might measure and test all proposals to cure, change or improve man. One could then say what is "human" and what is "inhuman." But I am not encouraged by the history of such attempts. Descriptive definitions of man are difficult enough; the most successful seem to me to be those which, rejecting attempts to find one necessary and sufficient description, move in the direction of combining a number of characteristics conjunctively. But the latter strategy, for all its value, can still leave a feeling of dissatisfaction. One is left with a clutter of valid descriptions, but in no way joined in an integrated way. Normative descriptions pose even worse problems, at the very least because any normative description must involve a procedure for deciding what to do with the data provided by descriptive definitions; and no descriptive definition tells one that. To know that man is a rational animal does not tell us, when a decision to act is called for, of what rational behavior should consist; the same can be said of any other definition of "man" or "human."

Philosophically, this is the old issue of how a move is to be made from the *is* to the *ought*. Nonetheless, by one means or another the move is made, but it is made, I want to argue, not by a series of logical steps but by the appropriation of images of man which try to do justice to man as he is and man as he might be. So long as it is man who must create the images, and link the descriptive and the normative, man is ineluctably a self-defining and self-imaging animal. Part of the self-definition will consist of finding a way of coping with the relationship between man as he is and man as one would like him to be.

Our own culture has moved very uneasily within this dialectic. At times, under the sway of the social and psychological sciences, Western culture has been tempted to shape its ideals from what people actually do, to reduce the gap between ideal and reality. Sometimes this is done in the name of greater "realism," and sometimes in the name of "honesty" or "authenticity." Unfortunately, this can be accomplished only by means of a considerable covert selectivity. The problem is that much of what people actually do is patently destructive; they harm and kill themselves and others. One is forced to hope for something better, at least in some respects. At that point, ideals which run counter to the empirical givenness of human behavior will and must be invoked. At quite the other extreme, images of man which work at the level of pure utopianism (of a romanticized past or idealized future) usually run aground on the realities of human behavior and the facticity of contemporary cultures.

Earlier I suggested that a drive for transcendence is a key

human characteristic. Man has always sought to become something other than he finds himself to be. He is forced to project images of future states, images of correction and improvement. Even if one says that such images should be grounded in what man essentially is—and that one should, in the first place, work to find that out—I do not see how even that much can be done without some provisional imaginative vision of a "better man" or an "improved" human nature to guide us in the search. Moreover, in our world any attempt to test a provisional vision will have to run the risk of error in the quest for knowledge and the control of nature. Abraham Edel has written some pertinent words in that respect:

> The domain of ignorance is and will be indefinitely vast. But from a practical point of view it can no longer be used as an *a priori* veto on attempts at knowledge and control. In more stable days, it could be said that no experimental ventures should be made in human life which involved a plunge into the unknown, because disasters might result. Now the same argument often can be urged against *not* making experimental ventures; for the consequences of continuing in the old ways in a rapidly changing world may be quite as unknowable and quite as disastrous. This argument does not justify recklessness in experiment; we are learning how reckless we have been. But it underlines the recklessness of conservatism too. In short, the emphasis falls on responsible attempts at control.[4]

Edel does not tell us what "responsible attempts at control" means, and that is the key question. But I think he is quite correct in arguing that to do nothing is, in a state of ignorance, as much a gamble as doing something, particularly at the present moment. For wittingly or unwittingly, any decision not to act will have to rest on some image of man, in this case involving a presumption that the human life we now know is at least as good as if not better than some hypothetical "improved" form of human life. That may be a valid point, but it is no less a utopian point than its converse.

In *The Abolition of Man,* C. S. Lewis wrote that "Each new power won *by* man is a power *over* man as well. Each advance leaves him weaker as well as stronger."[5] In the transition from pre-industrial to an industrial society this may have been true (though I must say I am skeptical). But we now live in an industrial-technological society and will continue to so live, as will our children. From that base point, Lewis' observation needs emendation. Each new power which man chooses deliberately not to win is the equivalent of an exercise of power over man as well. Each advance not made makes him weaker as well as stronger. I am not suggesting that the course of wisdom is to plunge on, to do everything that can be done; that is sheer stupidity. I am saying only that the general case for continuing attempts to improve human life and human beings is, in the face of uncertainty, as strong as the case for caution. This point does not, however, nullify Lewis' penetrating observation that the social meaning of technology has often been that of the development of new powers with which humans can control and harm each other; that still remains the greatest danger of technological advance.

When people are faced with general propositions of that kind, however, their image of man together with that of their image of a technological society is likely to suffuse and thus shape either their adventurousness or their caution. What, then, is a correct or sound image of man? It is that image which, on *sober* (not utopian) consideration, we find we can most creatively use in shaping the future. We can choose, and choose creatively, because the reality is rich enough in both positive and negative elements to leave us free to act in a way which can temper the hope for something better with prudent correctives provided by a reflection on past mistakes—but not for a minute in the process abandoning hope.

The word "creatively" lies at the heart of my argument. The failure of ethical codes which reduce demands on human beings by taking their statistically average conduct as a norm, making minute the distance between what people do and what they ought to do, rests on a static image of man, providing no space for growth or transcendence. No less static are attempts to fix, once and for all, an essence of man. Both the moral life and attempts at shaping an image of man should be likened to works of art, where a limited material reality is shaped by the imagination into a creation grounded in, but at the same time richer than, the reality from which it takes its beginnings. To ask the question "What is the normatively human?" is already to display a stirring of the imagination, looking for something which does not meet the naked eye, which is dissatisfied with received opinion and convention. The answer given to that question should respect its origins, that is, by working with images and concepts which are able to transmute the mundane into the stuff of realistic dreams and controlled strivings.

A number of issues need to be sorted out in any discussion of *the control of nature.* The first concerns the scope of the concepts of "nature" and "control." The second bears on the sense and extent to which nature is to be respected. The third concerns the relationship between man and nature, both descriptively and normatively.

I will not try to sort all of these issues out here. Instead, I want to suggest that most discussions of nature come to be dominated by several alternative models of the relationship between man and nature. The first I will call the *power and plasticity* model. In that model, nature is looked upon as essentially alien and independent of man, possessing no inherent value, dominated by impersonal forces and causes. For all these reasons, nature is seen as plastic, capable of being used, dominated and shaped as man sees fit. The human relationship to nature so understood is that it is man's unrestricted right to dominate, manipulate and control nature by the use of whatever power he can bring to bear in the service of his own goals. The only real limitations to the employment of power to subdue nature are those arising from a lack of knowledge of the workings of nature. That limitation imposes the need for a certain care and caution—nature can bite back—but the limitation is practical only, subject to being overcome by greater human ingenuity. The goal of knowledge is power, and any ethical problems which arise in the use of this power will be restricted solely to those bearing on the ends sought and means used by human beings; nature is not one of the ethical actors. Working with this model, death, for instance, might be seen as no more than one of nature's dirty tricks, to be overcome and outwitted by medical technology; death is the final absurdity.

Another common model may be called the *sacral and symbiotic*. In its religious forms, it sees nature as part of God's creation, to be respected and heeded. Man is not to be looked upon as the master of nature, but rather its steward; nature has been left in his trust. In its secular forms, this model sees man as a part of nature; if man is to be respected, so too should nature, which nurtures and sustains him. Nature has a value in and of itself, quite apart from man. Man should live in harmony and balance with nature, consciously seeing the relationship as symbiotic, even mutually rewarding. Under this model, nature is a teacher, showing us how to live with it and with ourselves. This model would be prone to see death as one of the rhythms of nature, to be gracefully accepted.

Still another model would neither sacralize nature nor reduce it to a mere object of manipulation. I will call it the *teleological* model, meaning by that a view of nature which sees a purposiveness and logic in nature, but without any necessary implication that there is a guiding hand behind those features. Thus it becomes possible to study nature and to discern its meaning and human significance. That knowledge can, in turn, give us considerable guidance in determining the wisest course of human conduct. At the very least, we can come to understand the possible folly in attempting to better nature, or to transcend its inherent limits. We are not necessarily bound by nature in any rigid sense, but we are at least able to see limitations on an unbridled intervention into nature. This model would also see death as one of the rhythms of nature, having its genetic and cultural place. There would be less of a tendency, however, to glorify death or those forces of nature which bring it on; it would induce at most a wise resignation in the face of death, as an unavoidable part of life.

There are many variants on these three models, but the details are less important in this case than the general contours and images. Each provides a way of responding differently to the problem of control. The power and plasticity model does, despite the Copernican revolution, keep man in the center of the universe, as its lord and master. Action, mastery and unceasing innovation become prized virtues. The sacral and symbiotic model put man and nature together in the center of the universe. Spontaneity, contemplation and passivity would be central virtues; control would not be sought nor manipulation pursued. The teleological model would not deny the need for some control. But it would seek a way of basing any need for control on an understanding of the inherent features of nature, letting them provide the main guidelines and limits.

Each of these models has important ethical implications.

The first is consistent with the position that man must be the creator of his own ethical norms, acting to bring order out of chaos by an act of unaided intelligence and will. There are no moral codes, absolutes or guides other than those man himself chooses to invent. The second model is supportive of an ethic derived entirely from a contemplation of the ways of nature; whatever is most in harmony with nature will be seen as the good. The third model would take nature as a partial guide, but with a realization that there is a difference between man and the rest of nature. Man must create his own ethical norms and moral codes, but in nature he can find the outline of a base for the foundation of the moral life.

In one form or another, some model of nature and man's relationship to nature stands behind every ethical system as well as specific ethical decisions bearing on technology. An ethical decision will show what a person takes man and nature to be about. That images of man and nature usually lie hidden well below the surface make them all the more potent. Moral hypocrisy is usually understood as a conscious dissimulation, in which people knowingly profess values they do not practice. Is it not possible, however, that while a person may genuinely be drawn to the ideals which he consciously professes, and thus be sincere enough about them, his actual behavior will be expressive of his subsurface images of the way reality actually is? Those images will determine what he actually does, forcing him to set aside that better self represented by what he professes. His "better self," which can articulate the way he would like the world to be, will in cases of conflict give way to the overwhelming power of those images of man and world which shape his real expectations. In the end, his real expectations will influence his sense of the extent to which he can afford, as the price of survival, to honor his vision of the world as he would like it to be. If moral behavior is to be changed, then, the place to begin is with images of man and the world; and that means moving beyond the realm of moral discourse.

As suggested earlier, the phrase "public morality" is not exactly a popular one in America. It conjures up pictures of the censor, the policeman, and the meddlesome, self-righteous moralist. The old saw that "you can't legislate morality" goes hand-in-hand with a repugnance toward any merging of law and ethics. How many times have we heard about the failure of prohibition, as the grand example of what happens when the state tries to impose values on people? The most that the law can do is enforce those standards of behavior which are required by the common good. The law does not tell people what is right and wrong in the ethical sense of those terms; it tells them only what they must do as the price of remaining citizens of reputable standing; that is, in order to stay out of jail.

That is an exceedingly naïve view of the law, one which the country has been forced to entertain because no satisfactory way has been found to combine the need for law with the fact of great differences on values. The simplest expedient has been to pretend that law and morals need have nothing to do with each other. But they obviously do. Laws reflect what the community values, what it believes essential for its common welfare. Are we really to believe that the civil rights legislation of the past two decades, or the abortion legislation and Supreme Court decision of the past few years, reflects only a shift in legal theory, having nothing to do with what people take to be right and wrong? Surely the prohibition of medical experimentation without the consent of those experimented upon says something about what the community believes one human being owes another. The words "freedom," "justice" and "equality" are not only specific legal concepts; their power as concepts derives from their ethical weight. Nor is it realistic to think that people's moral attitudes are not influenced by what the law permits and prohibits.

The great danger of the myth that law and morality should have nothing to do with each other does not lie only in the systematic distortion of reality it introduces. More importantly, it distracts the community from the necessity of trying, by public discussion, to shape some common values and moral standards which could pervade and enrich their life together. The expectation that one's neighbor will treat you with justice and respect cannot rest upon law alone. It must rest upon an expectation that one's neighbor privately cherishes the value of justice, that he would be just even if the law did not require it of him. Minority groups are well aware of the deficiencies of law as guarantors of respect and dignity. It is not just that laws may be evaded. More critically, law must have moral weight behind it. It must seem fitting and proper, consonant with and evocative of human hopes and aspirations for a livable community.

If in a general way the myth of a separation of law and morality is naïve and hazardous, it also has a particularly devastating effect in one of its versions. I am referring to that version which, popularly put, says that apart from what the law requires everyone should be free to shape his values and way of life as he sees fit. I honor the ideal of human freedom which this view represents, when seen as a political value. It bespeaks respect for conscience and the value of self-determination. Yet the privatizing of morality to which it can lead is too often apt to leave people utterly at sea about how they should shape their conscience and use their freedom. Many still have a religious tradition which will help them at that point; but an increasing number do not. Where, once people are free, are they to get their values? Moreover, unless people have values which are independent of the law, they will have no basis on which to judge law. The net result of a lack of common standards by which to make moral judgments may not, in the end, be that of increasing freedom at all, but of propagating herd behavior. A rapid fluctuation between worries about freedom and anxieties about survival may, in the end, be symptomatic of the absence of any real values at all. Popular opinion will hold sway. People will think they are free, but their behavior will be all too predictable, too much determined in actuality by class and status, or by fad and parochial tastes. Nothing more bedevils attempts to analyze and control technology than a widespread conviction in our society that ethical issues are not fit for rational, public discourse, that they are too personal, too subjective, too emotive to be subject to public scrutiny and judgment.

I do not know whether it will be possible to develop some common moral standards, or some common way of working through ethical dilemmas. But no effort seems to me more important. One function of a viable culture is that of providing the human beings who make it up with some common and sustaining values, some mutually rewarding ways of working together to achieve a more human life. It is certainly worth worrying about the possibility that any fresh attempts to develop the basis for a public morality will simply set the stage for a new social repression. But I am just as worried about the possibility that, in the absence of some well-thought-out standards which can be appropriated into private lives, there will exist no reasonable way at all for people to think through their ethical dilemmas. Nor, more generally, will there be any possibility of developing those deep habits of action, sustained by sane affective responses, which, as much or more than formal moral codes, establish the ethos of a culture. Quite apart from all of these motivations, there is the simple and painful fact that many of the hardest dilemmas posed by technology are insupportable and insoluble if they must be faced by individuals alone. A communal conscience will not be possible, and without such a conscience technology will both dominate and destroy.

## NOTES

1. Philip Rieff, "Fellow Teachers," *Salmagundi* No. 20 (Summer-Fall 1972), p. 19.
2. Philip Rieff, *The Triumph of the Therapeutic: Uses of Faith after Freud* (New York: Harper & Row, 1966), pp. 2–3.
3. *Ibid.*, p. 5.
4. Abraham Edel, "Scientists, Partisans and Social Control," *Transaction* (January 1972), p. 34.
5. C. S. Lewis, *The Abolition of Man* (New York: Macmillan, 1965), p. 71.

# CHAPTER 7

# *Population Control:*
## *SCIENTIFIC AND HUMAN LIMITS*

AS A CASE STUDY in the science of (technological) limits, excessive population growth and the need for its control presents a number of provocative features. Together with the case of DDT, it provides as perfect an example as one can find of an utterly unforeseen disaster resulting from an attempt to achieve a goal universally accepted as beneficial. It is quite arguable that the proliferation of automobiles (bringing attendant pollution problems) was in the first place a dubious goal, expressive mainly of the economic and social values of technological cultures. No such case could be made in the instance of goals which sought the control of death (the source of the population problem) or the control of malaria (the source of the DDT problem); those aims were accepted in all cultures, not reflecting any parochial cultural bias. In both instances, given the state of existing knowledge, there was every reason to assume that the chosen goals could be achieved without harmful, unintended consequences.

There is a particular ethical poignancy in the two cases. Even if it had been known that the introduction of DDT for the control of malaria would have had harmful environmental by-products, it is difficult to see (given the thousands of deaths annually from malaria) how any other choice could have been made at the time. Nor can one imagine how, had it been foreseen that the introduction of modern sanitation, higher standards of nutrition, better knowledge of prenatal needs, and the armamentarium of modern medicine would eventually create a population crisis, a conscious decision could have been made by any government or any people not to accept the danger of such a crisis. "Sufficient to the day is the evil thereof," is not just a biblical injunction; it is also one of the mandates of individual human survival. High infant mortality rates, controllable diseases (malaria, for instance), short life-spans, high maternal mortality figures, are intolerable human conditions, requiring immediate action even at the price of long-term evils.

There is, however, a limit to the DDT–population growth analogy. It is now possible to make use of new insecticides for the control of malaria which do not carry with them the environmental hazards of DDT; technological alternatives are possible, allowing the original goal to be sought with greater long-term safety. No such alternative has presented itself in the case of population growth. The lowering of deathrates as such produces population growth in the absence of a corresponding drop in birthrates. Given a universal unwillingness to return to high deathrates, the *only* available way to control population growth is by lowering birthrates. In a word, it is not the cause of population growth which can be controlled but only the effects.

This conclusion has some important implications for a science of limits. First, it presents a case where the need to limit the undesirable effects of technology cannot (or at least certainly will not) allow a return to that pretechnological state of affairs which saw population stability achieved by high deathrates; a culturally irreversible situation has been established by the development of modern medical strategies and technologies. If any "limit" to this technology is to be achieved (and population growth makes clear the need for a limit to be found and set), this will be possible culturally only by development of an effective birth-control technology.

Second, the drive to procreate has deep biological and cultural roots. As a species-phenomenon, it is the *sine qua non* of survival. As an individual phenomenon, it is a source of identity, the perpetuation of family and cultural lines, and in many traditional agricultural societies a requisite condition for family and group survival. Any social attempts, particularly of a precipitate sort, to intervene in and to control individual procreation will encounter strong resistance. If the existence of the problem in the first place indicates the presence of one kind of social pathology (unforeseen consequences of technological change), steps taken to deal with it could set the stage for another kind of social pathology—that which occurs when social measures are taken to control not the technology itself, but to control individual behavior in a way that will counteract the unwanted consequences of the technology. Population-control measures will not limit that medical technology which reduces deathrates, but that behavior—in this case, procreative—which makes the consequences of the death-control technology harmful. Inevitably, there will be a tendency to blame—and if blame does not work to coerce—those who refuse to alter their procreative behavior to meet the crisis.

Third, a further source of that social pathology which sees power brought to bear upon those thought to be behaving against the common good is the tacit assumption that, since they have not repudiated the control of death, they must be prepared to control birth. So irreversible is the cultural acceptance of death control that no parent would be allowed to ignore medical aid for a sick child. It is unimaginable that a couple would be allowed to strike a bargain which permitted them to ignore both death and birth control. They are now expected to accept both.

Population stability is possible only when there is a balance between births and deaths. That is the simple biological reality principle behind demographic statistics, a principle which cannot be violated without incurring either growth or diminishment of human populations. Since the latter is not at present the human problem, it is the former aspect of the principle which must serve as the limiting point. But there is a hard question to be faced. In trying to live with that point, how can that be done while taking account of the social reality principle pertinent to population growth and size, and of the pertinent psychological reality principle? I will define the social reality principle as the ineluctable tension between the pressure which will be felt by governments to limit growth, and the resistance which will be felt by the individual at attempts to control his or her individual procreation. And I will define the pertinent psychological reality principle as the following: while individuals will be willing to embrace the benefits of a control of death, they will not be as willing to give up the personal benefits of procreation, either for the good of the species or for the good of nations. Thus they will not, in their personal behavior, be willing to accept the biological reality principle; or better, they may be willing to accept it in theory but be unwilling to live it in their personal practice. Only one condition can change that situation: an acceptance of technological individualism, where the motives have little to do with the common good of the species or group.

A central paradox then emerges: the cure for the hazardous by-product of death control, population growth, will be—in the West—a cure which requires a deeper imbedding of that self-oriented individualism which is itself the source of the problems of Western technology in the first place. This paradox will have to be unraveled and an attempt made to see if a more powerful ethic than that of individualism can be brought to bear. This will require an extended attempt to grasp the ethical problems posed by population growth, understanding in particular some recent turns taken in the argument. One must begin with the population problem itself.

Throughout its history, the human species has been preoccupied with the conquest of nature and the control of death. Human beings have struggled to survive, as individuals, families, tribes, communities, and nations. Procreation has been an essential part of survival. Food could not have been grown, families sustained, individuals supported, or industry developed without an unceasing supply of new human beings. The result was the assigning of a high value to fertility. It was thought good to have children, for the children themselves, the parents, the society, and the species. While it may always have been granted that extenuating circumstances could militate against childbearing, the premise on which the value was based endured intact. There remained a presumptive right of individual procreation, a right thought to sustain the high value ascribed to the outcome: more human beings.

That the premise may now have to be changed, the value shifted, can only seem confounding. As Erik Erikson has emphasized, it is a risky venture to play with the "fire of creation," especially when the playing has implications for almost every aspect of individual and collective life.[1] The reasons for doing so would have to be grave. Yet excessive population growth presents such reasons—it poses critical dangers to the future of the species, the ecosystem, individual liberty and welfare, and the structure of social life. These hazards are serious enough to warrant a reexamination and, ultimately, a revision of the traditional value of unrestricted procreation and increase in population.

The main question is the way in which the revision is to proceed. If the old premise—the unlimited right of and need for procreation—is to be rejected or amended, what alternative premises are available? By what morally legitimate social and political processes, and in light of what values, are the possible alternatives to be evaluated and action taken? These are ethical questions, bearing on what is taken to constitute the good life, the range and source of human rights and obligations, the requirements of human justice and welfare. If the ethical problems of population limitation could be reduced to one overriding issue, matters would be simplified. They cannot. Procreation is so fundamental a human activity, so wide-ranging in its personal and social impact, that controlling it poses a wide range of ethical issues. My aim here is primarily to see what some of the different ethical issues are and to determine how an approach to them might be structured.

More pointedly, I will contend that those arguments which rest exclusively on the need for species survival are defective. But I will also argue that the right to individual procreation cannot be an unlimited or absolute personal right. The essential ethical issue of population control is to find a way to avoid both the tyranny of survival and the tyranny of individualism. The tyranny of survival would, if pressed far enough, open the way for a total suppression of all individual procreative choice. The tyranny of individualism would irrationally remove individual procreative behavior from the context of the common good. One could hardly find a better example of a situation where a public morality and a social ethic are needed, which could steer a safe course between the dangers of uncontrolled individual liberty or unrestricted government intervention. Only a different cultural super-ego can possibly deal with the population problem.

There would be no concern about population limitation if

there did not exist evidence that excessive population growth jeopardizes present and future welfare. Yet the way the evidence is evaluated will be the result of the values and interests brought to bear on the data. Every definition of the "population problem" or of "excessive population growth" will be value-laden, expressive of the ethical orientations of those who do the defining. While everyone might agree that starvation and malnutrition are bad, not everyone will agree that crowding, widespread urbanization, and a loss of primitive forest areas are equally bad. Human beings differ in their assessments of relative good and evil. To say that excessive population growth is bad is to imply that some other state of population growth would be good or better—for example, an "optimum level of population." But as the demographic discussion of an optimum has made clear, so many variables come into play that it may be possible to do no more than specify a direction: "the desirability of a lower *rate* [italics added] of growth."[2]

If the way in which the population problem is defined will reflect value orientations, these same definitions will have direct implications for the way in which the ethical issues are posed. An apocalyptic reading of the demographic data and projections—laying heavy stress on the dangers posed to "survival"—can, not surprisingly, lead to coercive proposals. Desperate problems are seen to require desperate and otherwise distasteful solutions.[3] Moreover, how the problem is defined, and how the different values perceived to be at stake are weighted, will have direct implications for the priority given to population problems in relation to other social problems. People might well agree that population growth is a serious issue, but they might (and often do) say that other issues are comparatively more serious.[4] If low priority is given to population problems, this is likely to affect the perception of the ethical issues at stake.

## *Population Control*

Excessive population growth raises ethical questions because it threatens existing or desired human values and ideas of what is good. In addition, all or some of the possible solutions to the problem have the potential for creating difficult ethical dilemmas. The decision to act or not to act in the face of the threats is an ethical decision. It is a way of affirming where the human good lies and the kinds of obligations individuals and societies have toward themselves and others. A choice in favor of action will, however, mean the weighing of different options, and most of the available options present ethical dilemmas.

In making ethical choices, decisions will need to be made on (1) the human good and values that need to be served or promoted—the ends; (2) the range of methods and actions consistent and coherent with those ends—the means; and (3) the procedure and rationale to be used in trying to decide both upon ends and means and upon their relation to each other in specific situations—the ethical criteria for decision-making. A failure to determine the ends, both ultimate and proximate, can make it difficult or impossible to choose the appropriate means. A failure to determine the range of possible means can make it difficult to serve the ends. A failure to specify or articulate the ethical criteria for decision-making can lead to capricious or self-serving choices, as well as placing obstacles in the way of a rational resolution of ethical conflicts.

In the case of ethics and the population problem, both the possibilities and the limitations of ethics become apparent. In the face of a variety of proposals to solve the population problem, some of them highly coercive, a sensitivity to the ethical issues and some greater rigor in dealing with them is imperative. The most fundamental matters of human life and welfare are at stake. Yet because of the complexity of the problem, including its variability from one nation or geographical region to the next, few hard and fast rules can be laid down about what to do in a given place at a given time. Still, since some choices must be made (and not to choose is to make a choice as well), the practical ethical task will be that of deciding upon the available options. While I will focus on some of the proposed options for reducing birthrates, they are not the only ones possible. Ralph Potter has discussed some others:

> It has generally been assumed that policy must be primarily, if not exclusively, concerned with bringing about a decline in the rate of population increase through a reduction in the birthrate. But there are other choices. It is generally considered desirable but impossible to increase resources at a sufficient pace and through an adequate duration to preserve the present level of living for all within an expanding population. It is generally considered possible but undesirable to omit the requirement that all persons have access to that which is necessary for a good life. There is still the option of redefining what is to be considered necessary for a good life or of foregoing some things necessary for a good life in order to obtain an equitable distribution in a society that preserves the autonomy of parents to determine the size of their families.[5]

A useful way of posing the issue practically is to envision the ethical options ranked on a preferential scale, from the most desirable to the least desirable. For working purposes, I will adopt as my own the formulation of Kenneth E. Boulding:–"A moral, or ethical, proposition is a statement about a rank order of preferences among alternatives, which is intended to apply to more than one person."[6] Ethics enters at that point when the preferences are postulated to have a value that transcends individual tastes or inclinations. Implicitly or explicitly, a choice among possibilities becomes an ethical decision when it is claimed that one or another alternative *ought* to be chosen—not just by me, but by others as well. This is where ethics differs from tastes or personal likings, which, by definition, imply nonobligatory preferences that are applicable to no more than one person (even if the tastes are shared).

I will assume at the outset that there is a problem of excessive population growth, a problem serious for the world as a whole (with a 2 percent annual growth rate), grave for many developing nations (where the growth rate approaches 3 percent per annum), and possibly harmful for the developed nations as well (with an average 1 percent growth rate). At the present rate of growth, world population will double in thirty-seven years. In many developing nations it will double in less than twenty-five years. The threats posed by excessive population growth are numerous: economic, environmental, agricultural, political, and socio-psychological. There is considerable agreement that something must be done to meet these threats. For the purpose of ethical analysis, the first question to be asked is, "In trying to meet these threats, what human ends are we seeking to serve?" Two kinds of human ends can be distinguished—proximate and ultimate.

Among the important proximate ends being sought in attempts to reduce birthrates in the developing countries are a raising of literacy rates, a reduction in dependency ratios, the elimination of starvation and malnutrition, more rapid economic development, and an improvement in health and welfare services; among the proximate ends in the developed countries are a maintenance or improvement of the quality of life, the protection of nonrenewable resources, and the control of environmental pollution. For most purposes, it will be sufficient to cite goals of that sort. But for ethical purposes it is critical to consider not just proximate but ultimate ends as well. For it is legitimate to ask of the specified proximate ends what ultimate human ends they are meant to serve. Why is it important to raise the literacy rate? Why is it necessary to protect nonrenewable resources? Why ought the elimination of starvation and malnutrition to be sought? For the most part, these are questions that need not be asked or that require no elaborate answers. Yet the ethical importance of such questions is that they force us to confront the goals of human life. Unless these goals are confronted at some point, ethics cannot start or finish.

Philosophically, solving the population problem can be viewed as determining at the outset what final values should be pursued. The reason, presumably, that a reduction in illiteracy rates is sought is that it is thought valuable for human beings to possess the means of achieving knowledge. The elimination of starvation and malnutrition is sought because of the self-evident fact that human beings must eat to survive. The preservation of nonrenewable resources is necessary in order that human life may continue through future generations. There is little argument about the validity of these propositions, because they all presuppose some important human values: knowledge, life and survival, for instance. Historically, philosophers have attempted to specify what, in the sense of "the good," human beings essentially seek. What do they, in the end, finally value? The historical list of values is long: life, pleasure, happiness, knowledge, freedom, justice, and self-expression, among others.

This is not the place to enter into a discussion of all of these values and the philosophical history of attempts to specify and rank them. Suffice it to say that four values have had a predominant role, at least in the West: freedom, justice, the general welfare, and security-survival. Many of the major ethical dilemmas posed by the need for population limitation can be reduced to ranking and interpreting these four values. Freedom is prized because it is a condition for self-determination and the achievement of knowledge. Justice, particularly distributive justice, is prized because it entails equality of treatment and opportunity and an equitable access to those resources and opportunities necessary for human development. The general welfare, encompassing adequate standards of living, education, literacy, beauty, and the like is a necessary ingredient of any life which is to go beyond mere physical survival. Security-survival is prized because it constitutes a fundamental ground for all human activities.

Excessive population growth poses ethical dilemmas because it forces us to weight and rank these values in trying to find solutions. How much procreative freedom, if any, should be given up in order to ensure the security-survival of a nation or a community? To ensure the general welfare? How much security-survival can be risked in order to promote distributive justice? How much procreative freedom can be tolerated if it jeopardizes distributive justice?

Ethical dilemmas might be minimized socially if there were agreement on the way the values ought to be ranked. One could say that freedom is so supreme a value that both justice and security-survival should be sacrificed to maintain it. But there are inherent difficulties in taking such a position. It is easily possible to imagine situations in which a failure to give due weight to the other values could result in an undermining of the possibility of freedom itself. If people cannot survive at the physical level, it becomes impossible for them to exercise freedom of choice, procreative or otherwise. If the freedom of some is unjustly achieved at the expense of the freedom of others, then the overall benefits of freedom are not maximized. If security-survival were given the place of supremacy, situations could arise in which this value was used to justify the suppression of freedom or the perpetuation of social injustice. In that case, those suppressed might well ask, "Why live, if one cannot have freedom and justice?"

For all of these reasons it is difficult and perhaps unwise to specify a fixed and abstract rank order of preference among the values: in some circumstances, each can enter a valid claim against the others. At the level of abstractions, all four values are critical; none can permanently be set aside. I have, moreover, greatly simplified the problem of values. A full analysis would have to encompass values in addition to those specified, i.e., cultural values of nations or of particular ethnic, racial or religious groups. Most of these values, at least for particular groups, could not be set aside either.

In the area of family planning and population limitation, a number of national and international declarations have given primacy to individual freedom. The Declaration of the 1968 United Nations International Conference on Human Rights is representative: ". . . couples have a basic human right to decide freely and responsibly on the number and spacing of their children and a right to adequate education and information in this respect."[7] While this primacy of individual freedom has been challenged,[8] it retains its position, serving as the ethical and political foundation of both domestic and foreign family-planning and population policies. Accordingly, it will be argued here that (*a*) the burden of proof for proposals to limit freedom of choice (whether on the grounds of justice or security-survival) rests with those who make the proposals, but that (*b*) this burden can, under specified conditions, be discharged if it can be shown that a limitation of freedom of choice in the name of justice or security-survival would tend to maximize human welfare and human values. This is only to say that, while the present international rank order of preference gives individual freedom primacy, it is possible to imagine circumstances that would require a revision of the ranking.

One way of approaching the normative issues of ranking preferences in population-limitation programs and proposals is by locating the key ethical actors, those who can be said to have obligations. Three groups of actors can be identified: individuals (persons, couples, families), the officers and agents of voluntary (private-external) organizations, and the government officials responsible for population and family-planning programs. I will limit my discussion here to individuals and governments. What are the ethical obligations of each of the actors? What is the right or correct course of conduct for them? I will approach these questions by first trying to define some general rights and obligations for each set of actors and then by offering some suggested resolutions of a number of specific issues.

I begin with individuals (persons, couples, families) because, in the ranking of values, individual freedom of choice has been accorded primacy by some international forums—and it is individuals who procreate. What are the rights and obligations of individuals with regard to procreation? Individuals have the right voluntarily to control their own fertility in accordance with their personal preferences and convictions.[9] This right logically extends to a choice of methods to achieve the desired control and the right to the fullest possible knowledge of available methods and their consequences (medical, social, economic and demographic, among others).

Individuals are obligated to care for the needs and respect the rights of their existing children (intellectual, emotional and physical); in their decision to have a child (or another child), they must determine if they will be able to care for the needs and respect the rights of the child-to-be. Since individuals are obliged to respect the rights of others, they are obliged to act in such a way that these rights are not jeopardized. In determining family size, this means that they must exercise their own freedom of choice in such a way that they do not curtail the freedom of others. They are obliged, in short, to respect the requirements of the common good in their exercise of free choice.[10] The source of these obligations is the rights of others.

The role of governments in promoting the welfare of their citizens has long been recognized. It is only fairly recently, however, that governments have taken a leading role in an antinatalist control of fertility.[11] This has come about by the establishment, in a number of countries, of national family-planning programs and national population policies. While many countries still do not have such policies, few international objections have been raised against the right of nations to develop them. So far, most government population policies have rested upon and been justified in terms of an extension of freedom of choice. Increasingly, though, it is being recognized that, since demographic trends can significantly affect national welfare, it is within the right of nations to adopt policies designed to reduce birthrates and slow population growth.

A preliminary question must, therefore, be asked. Is there any special reason to presume or suspect that governmental intervention in the area of individual procreation and national fertility patterns raises problems which, *in kind,* are significantly different from other kinds of interventions? To put the question another way, can the ethicopolitical problems that arise in this area be handled by historical and traditional principles of political ethics, or must an entirely new ethic be devised?

I can see no special reason to think that the formation of interventionist, antinatalist, national population policies poses any unique theoretical difficulties. To be sure, the perceived need to reduce population growth is historically new; there exists no developed political or ethicopolitical tradition dealing with this specific problem. Yet the principle of governmental intervention in procreation-related behavior has a long historical precedent: in earlier pronatalist population policies, in the legal regulation of marriage, and in laws designed to regulate sexual behavior. It seems a safe generalization to say that governments have felt (and generally been given) as much right to intervene in this area as in any other where individual and collective welfare appears to be at stake. That new forms of intervention may seem to be called for or may be proposed (that is, in an anti- rather than pro-natalist direction) does not mean that a new ethical or political principle is at issue. At least no such principle is immediately evident.

Yet if it is possible to agree that no new principles are involved, it is still possible to argue that a further extension of an old principle—the right of government intervention into procreation-related behavior—would be wrong. Indeed, it is a historical irony that, after a long international struggle to establish individuals' freedom of choice in controlling their own fertility, that freedom should immediately be challenged in the name of the population crisis. Irony or not, there is no cause to be surprised by such a course of events. The history of human liberty is studded with instances in which, for a variety of reasons, it has been possible to say that liberty is a vital human good and yet that, for the sake of other goods, restriction of liberty seems required. A classical argument for the need of a government is that a formal and public apparatus is necessary to regulate the exercise of individual liberty for the sake of the common good. And of course behind this reality lurks the Freudian perception of a permanent tension between the individual and society.

In any case, the premise of my discussion will be that governments have as much right to intervene in procreation-related behavior as in other areas of behavior affecting the general welfare. This right extends to the control of fertility in general and to the control of individual fertility in particular. The critical issue is the way in which this right is to be exercised—its conditions and limits—and that issue can be approached only by first noting some general issues bearing on the restriction of individual freedom of choice by governments.

Governments have the right to take those steps necessary to ensure the preservation and promotion of the common good—the protection and advancement of the right to life, liberty and property. The maintenance of an orderly and just political and legal system, the maintenance of internal and external security, and an equitable distribution of goods and resources are also encompassed within its rights. Its obligations are to act in the interests of the people, to observe human rights, to respect national values and traditions, and to guarantee justice and equality. Since excessive population growth can touch upon all of these elements of national life, responses to population problems will encompass both the rights and the obligations of governments. However, governmental acts should represent collective national decisions and be subject to a number of stipulations.

Let me recapitulate the points made so far and summarize some propositions, which I will then use to suggest solutions to some specific ethical issues.

*General moral rules:* (*a*) individuals have the right to freedom of procreative choice, and they have the obligation to respect the freedom of others and the requirements of the common good; (*b*) governments have the right to take those steps necessary to secure a maximization of freedom, justice, the general welfare and security-survival, and they have the obligation to act in such a way that freedom and justice are protected and welfare and security-survival enhanced.

*Criteria for ethical decision-making:* (*a*) one (individual, government, organization) is obliged to act in such a way that the fundamental values are respected; (*b*) in cases of conflict, one is obliged to act in such a way that any limitation of one or more of the fundamental values—a making of exceptions to the rules concerning these values—continues to respect the values and can be justified by the promise of increasing the balance of good over evil.

*Rank order of preference:* (*a*) the choices of action to be preferred are those that accord primacy to freedom of choice; (*b*) if conditions appear to require a limitation of freedom, this should be done in such a way that the direct and indirect harmful consequences are minimized and the chosen means of limitation are just—the less the harm, the higher the ranking.

Since it has already been contended that individual freedom of choice has primacy, the ethical issues to be specified here will concentrate on those posed for governments. This focus will, in any event, serve to test the limits of individual freedom.

Faced with an excessive population growth, several courses are open to governments. They can do nothing at all. They can institute, develop, or expand voluntary family planning programs—where the goal is individual welfare only or that of families. They can attempt to implement proposals that go beyond the establishment of voluntary family-planning programs to those which place the emphasis on population control by use of means which can, in some cases, severely limit or preclude freedom of choice. Bernard Berelson has listed some twenty-nine such proposed programs in his now classic article "Beyond Family Planning."[12]

Would it be right for governments to go beyond family planning if excessive population growth could be shown to be a grave problem? This question conceals a great range of issues. Who would decide if governments have this right? Of all the possible ways of going beyond family planning, which could be most easily justified and which would be the hardest to justify? To what extent would the problem have to be shown to be grave? As a general proposition, it is possible ethically to say that governments would have the right to go beyond family planning. The obligation of governments to protect fundamental values could require that they set aside the primacy of individual freedom in order to protect justice and security-survival. But everything would depend on the way they proposed to do so.

Would it be right for governments to establish involuntary fertility controls? These might include (if technically feasible) the use of a mass "fertility control agent," the licensing of the right to have children, compulsory temporary or permanent sterilization, or compulsory abortion.[13] Proposals of this kind have been put forth primarily as "last resort" methods, often in the context that human survival may be at stake. "Compulsory control of family size is an unpalatable idea to many," the Ehrlichs have written, "but the alternatives may be much more horrifying. . . . human survival seems certain to require population control programs."[14] Their own speculation is manifestly coercive: "If . . . relatively uncoercive laws should fail to bring the birthrate under control, laws could be written that would make the bearing of a third child illegal and that would require an abortion to terminate all such pregnancies."[15]

That last suggestion requires examination. Let us assume for the moment that the scientific case has been made that survival itself is at stake and that the administrative and enforcement problems admit of a solution. Even so, some basic ethical issues would remain. "No one," the United Nations has declared, "shall be subjected to torture or to cruel, inhuman, or degrading treatment or punishment."[16] It is hard to see how compulsory abortion, requiring governmental invasion of a woman's body, could fail to qualify as inhuman or degrading punishment. Moreover, it is difficult to see how this kind of suggestion can be said to respect in any way the values of freedom and justice. It removes free choice altogether, and in its provision for an abortion of the third child makes no room for distributive justice at all, since its burden would probably fall upon the poorest and least educated. It makes security-survival the prime value, but to such an extent and in such a way that the other values are ignored altogether.

But could not one say, when survival itself is at stake, that this method would increase the balance of good over evil? The case would not be easy to make (1) because survival is not the only human value at stake; (2) because the social consequences of such a law could be highly destructive (for example, the inevitable massive fear and anxiety about third pregnancies that would result from such a law); (3) because it would be almost impossible to show that this is the only method that would or could work to achieve the desired reduction in birthrates.

Would it be right for governments to develop "positive" incentive programs, designed to provide people with money or goods in return for a regulation of their fertility? These programs might include financial rewards for sterilization, for the use of contraceptives, for periods of nonpregnancy or nonbirth, and for family-planning bonds or "responsibility prizes."[17] In principle, incentive schemes are noncoercive; that is, people are not forced to take advantage of the incentive. Instead, the point of an incentive is to give them a choice they did not previously have.

Yet there are a number of ethical questions about incentive plans. To whom would they appeal most? Presumably, their greatest appeal would be to the poor, those who want or need the money or goods offered by an incentive program; they would hold little appeal for the affluent, who already have these things. Yet if the poor desperately need the money or goods offered by the incentive plan, it is questionable whether, in any real sense, they have a free choice. Their material needs may make the incentive seem coercive to them.

Thus, if it is only or mainly the poor who would find the inducements of an incentive plan attractive, a question of distributive justice is raised. Because of their needs, the poor have less choice than the rich about accepting or rejecting the incentive; this could be seen as a form of exploitation of poverty. In sum, one can ask whether incentive schemes are or could be covertly coercive, and whether they are or could be unjust.[18] If so, then while they may serve the need for security-survival, they may do so at the expense of freedom and justice. Would not the stage be set for some serious social pathologies?

At least three responses seem possible. First, if the need for security-survival is desperate, incentive schemes might well appear to be the lesser evil, compared with more overtly coercive alternatives. Second, the possible objections to incentive schemes could be reduced if, in addition to reducing births, they provided other benefits as well. For instance, a "family-planning bond" program would provide the additional benefit of old-age security.[19] Any one of the programs might be defended on the grounds that those who take advantage of it actually want to control births in any case (if this can be shown). Third, much could depend upon the size of the incentive benefit. At present, most incentive programs offer comparatively small rewards; one may doubt that they pose great dilemmas for individuals or induce significant psychological stress. The objections to such programs on the grounds of coercion become most pertinent if it can be shown that the recipients of an incentive benefit believe they have no real choice in the matter (because of their desperate poverty or the size of the benefit); so far, this does not appear to have been the case.[20]

While ethical objections have been leveled at incentive programs because of some experienced corrupt practices in their implementation, this seems to raise less serious theoretical issues. Every program run by governments is subject to corruption; but there are usually ways of minimizing it (by laws and review procedures, for instance). Corruption becomes a serious theoretical issue only when and if it can be shown that a government program is inherently likely to create a serious, inescapable, and socially damaging system of corruption. This does not appear to be the case with those incentive programs so far employed or proposed.

Would it be right for governments to institute "negative" incentive programs? These could take the form of a withdrawal of child or family allowances after a given number of children, a withdrawal of maternity benefits after a given number, or a reversal of tax benefits, to favor those with small families.[21] A number of objections to such programs have been raised. They are directly coercive in that they deprive people of free choice about how many children they will have by imposing a penalty on excess procreation—"excess" as determined by government and not by the individuals personally involved; thus they do not attach importance to freedom of choice. They can also violate the demands of justice, especially in those cases where the burden of the penalties would fall upon those children who would lose benefits available to their siblings. And the penalties would probably be more onerous to the poor than to the rich, further increasing the injustice. Finally, from quite a different perspective, the social consequences of such programs might be most undesirable. They could, for instance, worsen the health and welfare of those mothers, families and children who would lose needed social and welfare benefits. Moreover, such programs could be patently unjust in those places where effective contraceptives do not exist (most places at present). In such cases, people would be penalized for having children whom they could not prevent with the available birth-control methods.

It is possible to imagine ways of reducing the force of these objections. If the penalties were quite mild, more symbolic than actual (as Garrett Hardin has proposed),[22] the objection from the viewpoint of free choice would be less; the same would apply to the objection from the viewpoint of justice. Moreover, if the penalty system was devised in such a way that the welfare of children and families would not be harmed, the dangerous social consequences would be mitigated. Much would depend, in short, upon the actual provisions of the penalty plan and the extent to which it could minimize injustice and harmful social consequences.

Nonetheless, penalty schemes raise serious ethical problems. They would be justifiable only if it could be shown that survival or the general welfare was at stake and that, in their application, they would give due respect to freedom and justice. Finally, it would have to be shown that, despite their disadvantages, they promised to increase the balance of good over evil—which would include a calculation of the harm done to freedom and justice and a weighing of other, possibly harmful, social consequences. If these are serious concerns in the case of a threat to survival, they would be even more serious if they were used for the sake of improving the general welfare. Quite clearly some individuals would be forced to make major personal sacrifices for only small incremental gains for the populace as a whole.

An additional problem should be noted. Any penalty or benefit scheme would require some method of governmental surveillance and enforcement. Penalty plans, in particular, would invite evasion—for example, hiding the birth of children to avoid the sanctions of the scheme. This likelihood would be enhanced among those who objected to the plan on moral or other grounds, or who believed that the extra children were necessary for their own welfare. One does not have to be an ideological opponent of "big government" to imagine the difficulties of trying to ferret out violators or the lengths to which some couples might go to conceal pregnancies and births. Major invasions of privacy, implemented by a system of undercover agents, informants, and the like, would probably be required to make the scheme work. To be sure, there are precedents for activities of this kind (as in the enforcement of income tax laws), but the introduction of further governmental interventions of this kind would raise serious ethical problems, creating additional strains on the relationship between the government and the people. The ethical cost of an effective penalty system would have to be a key consideration in the development of any penalty program. "Ethical costs," it might be said, could at the very least (quite apart from ethics as such) be considered a cost to the general welfare.

Would it be right for governments to introduce antinatalist shifts to social and economic institutions? Among such shifts might be a raising of marriage ages, manipulation of the family structure away from nuclear families and bonuses for delayed marriage.[23] The premise of these proposals is that fertility patterns are influenced by the context in which choices are made and that some contexts (for example, higher female employment) are anti- rather than pro-natalist. Thus, instead of intervening directly into the choices women make, these proposals would alter the environment of choice; freedom of individual choice would remain. The attractiveness of these proposals lies in their putative noninterference with choice. They do not seem to involve coercion. But they are not without their ethical problems, at least in some circumstances. A too-heavy weighting of the environment of choice in an antinatalist direction would be tantamount to an interference with freedom of choice. An environment which made childbearing difficult or burdensome would hardly provide as free a setting for choice as one where childbearing was easy. In some situations, a manipulation of the institution of marriage (for example, raising the marriage age) could be unjust, especially if there exist no other social options for women.

The most serious problems, however, lie in the potential social consequences of changes in basic social institutions. What would be the long-term consequences of a radical manipulation of family structure for male-female relationships, for the welfare of children, for the family? One might say that the consequences would be good or bad, but the important point is that they would have to be weighed. Should some of them appear bad, they would then have to be justified as entailing a lesser evil than the continuation of high birthrates. If some of the changes promised to be all but irreversible once introduced, the justification would have to be all the greater. However, if the introduction of shifts in social institutions had some advantages in addition to antinatalism—for instance, greater freedom for women, a value in its own right—these could be taken as offsetting some other, possibly harmful, consequences.

Would it be right for the government of a developed nation to make the establishment of a population-control program in a developing nation a condition for extending food aid?[24] This would be extremely difficult to justify on ethical grounds. At the very least, it would constitute an interference in a nation's right to self-determination.[25] Even more serious, it would be a direct exploitation of one nation's poverty in the interests of another nation's concept of what is good for it; and that would be unjust. Finally, I would argue that, on the basis of Article 3 of the "Universal Declaration of Human Rights,"[26] a failure to provide needed food aid would be a fundamental violation of the right to life (when that aid could, without great cost to the benefactor nation, be given). The argument that such aid, without an attendant population-control program, would only make the problem worse in the long run, is defective. Those already alive, and in need of food, have a right to survival. To willfully allow them to die, or to deprive them of the necessities of life, in the name of saving even more lives at a later date cannot be justified in the name of a greater preponderance of good over evil. There could be no guarantee that those future lives would be saved, and there would be such a violation of the rights of the living (including the right to life) that fundamental human values would be sacrificed. That proposal is an exemplary example of the tyranny of survival.

Would it be right for a government to institute programs that go beyond family planning—particularly in a coercive direction—for the sake of future generations? This is a particularly difficult question, in great part because the rights of unborn generations have never been philosophically, legally, or ethically analyzed in any great depth.[27] On the one hand, it is evident that the actions of one generation can have profound effects on the options available to future generations. And just as those living owe much of their own welfare to those who preceded them (beginning with their parents), so too the living would seem to have obligations to the unborn. On the other hand, though, the living themselves have rights—not just potential, but actual. To set aside these rights, necessary for the dignity of the living, in favor of those not yet living would be to act arbitrarily.

A general solution might be suggested. While the rights of the living should take precedence over the rights of unborn generations, the living have an obligation to refrain from actions that would endanger future generations' enjoyment of the same rights that the living now enjoy. This means, for instance, that the present generation should not exhaust non-renewable resources, irrevocably pollute the environment, or procreate to such an extent that future generations will be left with an unmanageably large number of people. All of these obligations imply a restriction of freedom. However, since the present generation does have the right to make use of natural resources and to procreate, it must be demonstrated (not just asserted) that the conduct of the present generation poses a direct threat to the rights of future generations. In a word, the present generation cannot be deprived of rights on the basis of vague speculations about the future or uncertain projections into the future.

Do governments have the right unilaterally to introduce programs that go beyond family planning? It is doubtful that they do. Article 21 of the "Universal Declaration of Human Rights" asserts that "Everyone has the right to take part in the government of his country, directly or through freely chosen representatives. . . . The will of the people shall be the basis of the authority of government."[28] There is no evident reason that matters pertaining to fertility control should be exempt from the requirements of this right. By implication, not only measures that go beyond family planning, but family-planning programs as well require the sanctions of the will of the people and the participation of the people in important decisions.

The preceding list of specific issues by no means exhausts the range of possible ethical issues pertaining to governmental action; it is meant only to be illustrative of some of the major issues. Moreover, the suggested solutions are only illustrative. The complexities of specific situations could well lead to modification of them. That is why ethical analysis can rarely ever say exactly what ought to be done in *x* place at *y* time by *z* people. It can suggest general guidelines only.

I want now to propose some general ethical guidelines for governmental action ranking from the most preferable to the least preferable.

1. Given the primacy accorded freedom of choice, governments have an obligation to do everything within their power to protect, enhance and implement freedom of choice in family planning. This means the establishment, as the first

order of business, of effective voluntary family-planning programs.

2. If it turns out that voluntary programs are not effective in reducing excessive population growth, then governments have the right, as the next step, to introduce programs that go beyond family planning. However, in order to justify the introduction of such programs, it must be shown that voluntary methods have been adequately and fairly tried, and have nonetheless failed and promise to continue to fail. It is highly doubtful that, at present, such programs have "failed"; they have not been tried in any massive and systematic way.[29]

3. In choosing among possible programs that go beyond family planning, governments must first try those which, comparatively, most respect freedom of choice (that is, are least coercive). For instance, they should try "positive" incentive programs and manipulation of social structures before resorting to "negative" incentive programs and involuntary fertility controls.

4. Further, if circumstances force a government to choose programs that are quasi or wholly coercive, they can justify such programs if, and only if, a number of prior conditions have been met: (*a*) if, in the light of the primacy of free choice, a government has discharged the burden of proof necessary to justify a limitation of free choice—and the burden of proof is on the government (this burden may be discharged by a demonstration that continued unrestricted liberty poses a direct threat to distributive justice or security-survival); and (*b*) if, in light of the right of citizens to take part in the government of their country, the proposed limitations on freedom promise, in the long run, to increase free choice, decisions to limit freedom are collective decisions, the limitations on freedom are legally regulated and the burden falls upon all equally, and the chosen means of limitation

respect human dignity, which will here be defined as respecting those rights specified in the United Nations "Universal Declaration of Human Rights." The end—even survival—does not justify the means when the means violate human dignity and logically contradict the end.

As a general rule, the more coercive the proposed plan, the more stringent should be the conditions necessary to justify and regulate the coercion. In addition, one must take account of the possible social consequences and potential social pathologies of different programs, consequences over and above their impact on freedom, justice, the general welfare and security-survival. Thus, if it appears that some degree of coercion is required, that policy or program should be chosen which entails the least amount of coercion, limits the coercion to the fewest possible cases, is most problem-specific, allows the most room for dissent of conscience, limits the coercion to the narrowest possible range of human rights, threatens human dignity least, establishes the fewest precedents for other forms of coercion and is most quickly reversible if conditions change.

While it is true to say that social, cultural and political life requires, and has always required, some degree of limitation of individual liberty—and thus some coercion—that precedent does not in itself automatically justify the introduction of new limitations.[30] Every proposal for a new limitation must be justified in its own terms—the specific form of the proposed limitation must be specifically justified. It must be proved that it represents the least possible coercion consonant with the required goal, that it minimizes injustice to the greatest extent possible, that it gives the greatest promise of enhancing security-survival, and that it has the fewest possible harmful consequences and potential social pathologies (both short- and long-term).

The approach I have taken to population limitation has been cautionary. I have accepted the primacy of freedom of choice as a given not only because of its primacy in United Nations and other declarations, but also because it is a primary human value. I have suggested that the burden of proof must lie with those proposals, policies or programs that would place the primacy elsewhere. At the same time, I have laid down numerous conditions necessary to discharge the burden of proof. Indeed, these conditions are so numerous, and the process of ethical justification so difficult, that the possibility of undertaking decisive action may seem to have been excluded. This is a reasonable concern, particularly if time is short. Is it reasonable to give the ethical advantage to freedom of choice?[31] Does this not mean that a great chance is being taken? Is it not unethical to take risks of that sort, and all the more so since others, rather than ourselves, will have to bear the burden if the risk-taking turns out disastrously? In particular, would it not be irresponsible for governments to take risks of this magnitude? Can governments gamble when survival may be at stake?

Three kinds of responses to these questions are possible. First, as mentioned, it can and has been argued that freedom of choice has not been adequately tested. The absence of a safe, effective and inexpensive contraceptive has been one hindrance, particularly in developing countries. It is reasonable to expect that such a contraceptive will eventually be developed. The weakness of existing family-planning programs (and population policies dependent upon them) has, in great part, been the result of inadequate financing, poor administration, and scanty research and survey data. These are correctable deficiencies, assuming that nations give population limitation a high priority. If they do not, it is unlikely that more drastic population policies could be successfully introduced or implemented. Very little effort has been expended anywhere in the world to educate people and persuade them to change their procreation habits. Until a full-scale effort has been made, there are few good grounds for asserting that voluntary limitation will be ineffective.

Second, while the question of scientific-medical-technological readiness, political viability, administrative feasibility, economic capability and assumed effectiveness of proposals that would go beyond family planning is not directly ethical in nature, it has important ethical implications. If all of these categories seem to militate against the practical possibility of instituting very strong, immediate or effective coercive measures, then it could become irresponsible to press for or support such measures. This would especially be the case if attention were diverted away from what could be done, for example, an intensification of family-planning programs.

Third, primacy has been given to freedom of choice for ethical reasons. Whether this freedom will work as a means of population limitation is a separate question. A strong indication that freedom of choice will be ineffective does not establish grounds for rejecting it. Only if it can be shown that the failure of this freedom to reduce population growth threatens other important human values, thus establishing a genuine conflict of values, would the way be open to remove it from the place of primacy. This is only another way of asserting that freedom of choice is a right, grounded in a commitment to human dignity. The concept of a "right" becomes meaningless if rights are wholly subject to tests of economic, social or demographic utility, to be given or withheld depending upon their effectiveness in serving social goals.

In this sense, to predicate human rights at all is to take a risk. It is to assert that the respect to be accorded human beings ought not to be dependent upon majority opinion, cost-benefit analysis, social utility, governmental magnanimity or popular opinion. While it is obviously necessary to adjudicate conflicts among rights, and often to limit one right in order to do justice to another, the pertinent calculus is that of rights, not of utility. A claim can be entered against the primacy of one right only in the name of one or more other important rights. The proper route to a limitation of rights is not directly from social facts (demographic, economic, and so on) to rights, as if these facts were enough in themselves to prove the case against a right. The proper route is to show that the social facts threaten rights, and in what way, and from there to show that a limitation of one right may be necessary to safeguard or enhance other rights. To give primacy to the right of free choice is to take a risk. The justification for the risk is the high value assigned to the right, a value that transcends utilitarian considerations.

Yet even if it is granted that the risk of freedom should be run, will it not contribute to the danger of a tyrannical individualism, which sets the good of individual procreative choice above that of communal justice, welfare and survival? Why should the latter values have to wage the risk of an uphill battle against individualism? In theory, there could be such a risk. But its presence would be possible only if it were assumed that procreation, the having of children, is to be lumped without distinction among all the other varieties of individual desires in a technological society. In such societies people want, in addition to children, more automobiles, highways, electric carving knives and can openers, larger and faster jet airplanes, television sets and air conditioners. Is the desire for children to be put on a level of parity with all these other items?

In many respects, just this kind of parity was established in the recent debate over United States population policy. The early impetus to that debate was provided in 1968 by Paul Ehrlich in *The Population Bomb,* which traced all the ills of our (technological) society to "too many people." It was not a thesis which could stand much rigorous investigation, and by the early 70s the counterattack of Barry Commoner and others had succeeded in achieving an uneasy stalemate between those who put the blame for social ills on uncontrolled population growth and those who put it on the uncontrolled growth of technology. Thus it now seems to be agreed, as a congenial consensus position, that both population and technological growth are together the responsible villains. The practical moral injunction which results is direct: thou shalt limit the size of thy family and restrict thy use of technology. This injunction makes it difficult to escape the conclusion that the right to children and the right to the benefits of technology are rights of a similar, and equally limited, value.

That is a strange outcome, particularly in technological societies where most of the desires spawned by modern technology did not, unlike the desire to procreate, even exist as short a time as one century ago. The right to have an automobile—which human beings managed to live without for thousands of years—has achieved an equality with the right to have children, a most remarkable transmutation of values. Individual couples are now told it is their duty to limit the size of their family in order to protect and sustain the achieved "quality of life," a life which now includes the automobile and a whole array of other technological items. In short, if too many people have too many children, then too many other people will not be able to have automobiles (because of pollution, highway crowding, etc.); thus family size should be limited, QED.

One must at this point come to grips with the paradox noted at the beginning of this chapter. The paradox comes out of the fact that the most culturally persuasive argument in favor of population control in technological societies is the threat which uncontrolled population growth will pose to individualism. "Survival" is a heavily invoked value; but the survival in question is the preservation of the technological way of life. The paradox can be put in the form of a hypothetical imperative: if you do not control your desire to have children, then you will be forced to control your desire to partake fully of the benefits which a technological society can contribute to your personal well-being. The most prominent arguments in favor of family planning and population control in technological societies center, not surprisingly, on the *individual* (not national) economic, personal and cultural gains which fewer children will bring.

In that respect, data on the economic cost to parents of raising children, the psychological cost to women seeking personal self-realization, the tax burden on individuals occasioned by a large percentage of children in a society, the possibilities of restrictions being placed on the number of automobiles one might be able to own and the number of visits to national parks one might be able to make—considerations of that kind have been prominent among the arguments advanced in favor of population control in the United States. Even if the motives of those advancing such highly individualistic rationales for population control stem from an ultimate concern for the common good, the fact that they are forced to cast their arguments in terms of individual, technological benefits is in itself a highly revealing commentary on the depth to which technological individualism has imbedded itself in the cultural super-ego of technological societies.

And there lies the greatest danger of government-imposed population-control measures. For the most likely motive for such measures in affluent societies would be to preserve the technological status quo, one which has historically been dedicated to the fulfillment of individual desires, whatever the nature of those desires and however unlimited their demands. With this kind of technological individualism at its core, imposed population-control policies could hardly fail to introduce a devastating social pathology. War would then have to be declared against those who, in the eyes of others, were seen as a threat to the fulfillment of private desires. Those who wanted an unrestricted right to own automobiles would have to wage a bitter struggle against the "breeders," those whose values lead them toward children rather than automobiles.

The most basic reason to be suspicious of Garrett Hardin's formula "mutual coercion mutually agreed upon," is *not* that it represents a break with the democratic tradition, or that it is incompatible with human dignity, or that it might amount to the repression of some groups in the society, or that the seriousness of the population problem does not warrant any coercion. All humane, democratic societies work on the principle of "mutual coercion mutually agreed upon," and as suggested earlier there is no reason in principle why the right to procreate should not be brought under some limitations if the situation required it. The question one has to ask is whether the coercion to be agreed upon would find its impetus in a basic concern for the preservation of central human rights threatened by population growth, or—quite a different matter—from a desire on the part of individuals to protect their technological prerogatives and ambitions. It is the latter which is to be feared, particularly when couched in the deceptive language of survival. For it is not "survival" at all which is at stake, but rather the preservation of technological individualism with its high base-line of psychological survival, a form of survival obsession which flourishes only in the richest, most advanced societies, having little or nothing to do with either minimal physical survival or even that adequate, modest form of survival which cultures were able to attain well before the advent of advanced technology.

Here is an instance where one might well hope that the optimism Freud expressed about the power of altruism would express itself. In that event, the motive for population control would be a concern for the welfare of the human community and human rights; it would seek to draw people closer in the face of the fact that the resources and space of the earth are not unlimited. Then an easily defensible case could be made for population limitation. As matters now stand—at least in the West—the movement for control has not been able to rid itself of the contamination of individualism, and thus the threat of a social pathology quite as hazardous in its implications as the growth of population itself. When the having of children becomes an obscenity, but the production and consumption of automobiles continues apace, something is fundamentally wrong: a social pathology is already present. When a science of limits is devised which focuses more of its attention on ways to limit people rather than ways to limit technology, there is reason to be fearful for the future. One must then hope—without many contemporary grounds for doing so—that the long-standing psychological reality principle which has seen children and human beings as more important than the technological good life will reassert itself.

## NOTES

1. Erik H. Erikson, *Insight and Responsibility* (New York: Norton, 1964), p. 132.
2. Bernard Berelson, in *Is There an Optimum Level of Population?*, ed. S. F. Singer (New York: McGraw-Hill, 1971), p. 305.
3. See, for instance, P. R. Ehrlich and A. H. Ehrlich, *Population, Resources, Environment: Issues in Human Ecology* (San Francisco: Freeman, 1970), pp. 321–24.
4. A 1967 Gallup Poll, for example, revealed that, while 54 percent of those surveyed felt that the rate of American population growth posed a serious problem, crime, racial discrimination, and poverty were thought to be comparatively more serious social problems. (J. F. Kanther, *Studies in Family Planning*, No. 30 [May 1968], p. 6.)
5. Ralph Potter, "Coercion and Population Policy," unpublished

paper prepared for the Institute of Society, Ethics and the Life Sciences.

6. Kenneth E. Boulding, *American Economics Review,* Vol. 59 (March 1969), p. 1.
7. *Final Act of the International Conference on Human Rights* (United Nations, New York, 1968), p. 15.
8. For instance, not only has Garrett Hardin, in response to "The World Leaders' Statement," denied the right of the family to choose family size, he has also said that "If we love the truth we must openly deny the validity of the Universal Declaration of Human Rights, even though it is promoted by the United Nations" ("The Tragedy of the Commons," *Science,* Vol. 162 [December 13, 1968], p. 246). How literally is one to take this statement? The declaration, after all, affirms such rights as life, liberty, dignity, equality, education, privacy and freedom of thought. Is none of these rights valid? Or, if those rights are to remain valid, why is only the freedom to control family size to be removed from the list?
9. *Final Act . . . op. cit.,* p. 15.
10. See A. S. Parkes, in *Biology and Ethics,* ed. F. J. Ebling (New York: Academic Press, 1969), pp. 109–16.
11. In general, "antinatalist" means "attitudes or policies directed toward a reduction of births," and "pronatalist" means "attitudes or policies directed toward an increase in birth."
12. See Bernard Berelson, "Beyond Family Planning," *Studies in Family Planning,* No. 38 (February 1969), p. 1.
13. *Ibid.*
14. Ehrlich and Ehrlich, *op. cit.,* p. 256.
15. *Ibid.,* p. 274.
16. "Universal Declaration of Human Rights," in *Human Rights: A Compilation of International Instruments of the United Nations* (United Nations, New York, 1967), Article 5.
17. Berelson, "Beyond Family Planning," *op. cit.,* p. 2.
18. See, for example, W. H. Davis, *The New Republic* (June 20, 1970), p. 19.
19. See Ronald G. Ridker, *Studies in Family Planning,* No. 43 (June 1969), p. 11.
20. The payments made in six different family-planning programs are listed in *Incentive Payments in Family Planning Pro-*

*grammes* (International Planned Parenthood Federation, London, 1969), pp. 8–9.

21. Berelson, *op. cit.*, p. 2.
22. Garrett Hardin, "Multiple Paths to Population Control," *Family Planning Perspectives*, Vol. 2 (June 1970), p. 26.
23. Berelson, *op. cit.*, pp. 2–3.
24. Paul R. Ehrlich, *The Population Bomb* (New York: Ballantine Books, 1968), pp. 158–73.
25. See the "International Covenant on Economic, Social and Cultural Rights," Article 1, section 1, paragraph 1, in *Human Rights: A Compilation of International Instruments of the United Nations* (United Nations, New York, 1967), p. 4: "All people have the right to self-determination. By virtue of that right they freely determine their political status and freely pursue their economic, social and cultural development."
26. "Everyone has the right to life, liberty and the security of person" ("Universal Declaration of Human Rights," *ibid.*).
27. Among the few recent discussions on the obligation to future generations are M. P. Golding, "Obligations to Future Generations," *Monist*, Vol. 56 (January 1972), pp. 85–99; and Daniel Callahan, "What Obligations Do We Have to Future Generations?" *American Ecclesiastical Review*, Vol. 164 (April 1971), pp. 265–80.
28. "Universal Declaration of Human Rights," *ibid.*
29. See D. Nortman, in *Reports on Population/Family Planning* (Population Council, New York, December 1969), pp. 1–48. Judith Blake is pessimistic about the possibilities of family-planning programs, "Population Policy for Americans: Is the Government Being Misled?" *Science*, Vol. 164 (May 2, 1969), pp. 522–29.
30. See E. Pohlman, *Eugenics Quarterly*, Vol. 13 (June 1966), p. 122: "The spectre of 'experts' monkeying around with such private matters as family size desires frightens many people as being too 'Big Brotherish.' But those involved in eugenics or psychotherapy, or child psychology, or almost any aspect of family planning are constantly open to the charge of interfering in private lives, so that the charge would not be new. . . . Of course, many injustices have been done with the rationale of being 'for their own good.' But the population avalanche may

be used to justify—perhaps rationalize—contemplation of large-scale attempts to manipulate family size desires, even rather stealthily." This mode of reasoning may explain how some people will think and act, but it does not constitute anything approaching an ethical justification.

31. Paul R. Ehrlich argues in *The Population Bomb* (*op. cit.,* pp. 197–98) that the taking of strong steps now to curb population growth is the wiser and safer gamble than doing nothing or too little. This seems to me a reasonable enough position, up to a point. That point would come when the proposed steps would seriously endanger human dignity; an ethic of survival, at the cost of other basic human values, is not worth the cost.

# CHAPTER 8

# *The Genetic Quality of Human Life*

IF POPULATION GROWTH presents a prototypical case of one class of problems introduced by technological advances, the reality of genetic counseling and the prospect of genetic engineering presents another, and different, one. Population growth stole upon the world unnoticed and unanticipated; action has had to be taken after the fact. Genetic counseling and engineering, by contrast, present a picture where most of the potentially major technological advances are still in the future. There still remains time to make fundamental decisions about future directions, to think through basic values, to determine and then judge possible consequences.

To be sure, a case can be made that the human community has long engaged in genetic manipulation, that is, in shaping the composition and quality of the human gene pool and in modifying individual genetic makeup. All societies have had laws or customs regulating marriage, particularly marriage among siblings and close relations; such laws had, and continue to have, a genetic impact. Moreover, there is some agreement that the interaction between genetic composition and environment can, over a long period of time, result in basic genetic change. No less decisively, the control of some genetically based diseases—diabetes, phenylketonuria (PKU), retinoblastoma, for example—means that these diseases may gradually encompass an ever-higher proportion of people. Those who carry them are enabled by modern medicine to live, reproduce and thus transmit (and have their children transmit) the disease in an ever-widening circle; the forces of natural selection are thereby circumvented. Each of these instances can be brought to bear in support of the argument that humans have long engaged in genetic manipulation, even if they were not aware of it.

Yet even if this argument is accepted, there can be little doubt that both the understanding of genetics and the technological capability of using this knowledge in clinical practice have increased so rapidly and dramatically that an entirely new situation presents itself. Amniocentesis—the medical technique which makes possible an analysis of the amniotic fluid of a pregnant woman and, thereby, the detection of some genetic diseases in the fetus she carries—makes it possible (in conjunction with liberal abortion laws) for parents to decide whether to continue a pregnancy or to have an abortion.[1] Here is a form of prior, or prenatal, knowledge which simply was not available even a few years ago. The imminent development of the fetoscope—which will allow visual scanning of a fetus together with the drawing of a fetal blood sample—will greatly expand the range of foreknowledge of the genetic outcome of individual pregnancies. The advent of mass genetic screening for the detection of the carriers of harmful genes or the detection of genetic disease will increasingly make it possible for individuals prior to marriage or prior to procreation to know the likelihood of their producing a genetically defective child. It will also become possible for society to identify those capable of transmitting genetic disease.[2]

All of those examples are drawn from the general field of genetic counseling. They are present or near-distant developments. Genetic engineering, encompassing gene replacement with normal DNA, cloning, *in vitro* fertilization, artificial placentas and the like are in the more distant future. Even now, however, the basic theoretical knowledge is being gathered and the primitive technologies being designed which could make all of these possibilities a reality in the next one to three decades. That probability makes it possible to do some advance thinking, not only about whether those developments will be good for human life but also, in the meantime, whether and to what extent the basic research which will make them possible should (*a*) continue at all, (*b*) be accelerated or (*c*) be continued or accelerated under conditions of public approval and monitoring. Even in the case of those technologies put in the service of genetic counseling, it is possible to ask the same kinds of questions.

No attempt will be made here to review all of the recent advances in genetics, genetic counseling and genetic engineering. Nor will an attempt be made to survey the whole range of ethical, legal and public policy questions which they raise.[3] My focus will be on the fundamental, underlying question of the extent to which it is ethically legitimate to attempt to control the genetic quality of human life; and, if so, to what extent and by what means. Toward that end, I will make some basic assumptions about the present state of cultural attitudes toward the control of genetic quality, each of which I believe is a reasonably safe assumption:

1. The warm public response to the increased availability of individual genetic counseling (including amniocentesis and prenatal diagnosis), the growing public acceptance of abortion in the case of known or suspected fetal abnormality, and the greater emphasis in technological, population-conscious societies on having few children and "normal" wanted children,

together mean the advent of an important and growing psychological reality principle: contemporary parents are less willing than in earlier generations to bear defective children, and thus strongly motivated to take active steps to avoid the possibility. Parents have, as individuals, a greater desire to control the genetic quality of their offspring. On the whole, they will welcome, not fear, the advent of those scientific and technological developments which increase the control. In addition, they are likely to welcome the possibility of mass screening for genetic disease (so long, *at present,* as the screening is not coercive or mandatory).

2. The great public acceptance of genetic counseling, and the growing demand for counseling services (at a time when the birthrate is low and continuing to drop), means that public and private support of research in the field is likely to continue and even to accelerate. This in turn will increase the probability of even more rapid scientific and technological advances in the field of genetics in the years ahead.

3. Although there now exists some small degree of public controversy over the ethical and social legitimacy of research and applied technology in the area of genetic counseling,[4] there is no evidence that the existing opposition is strong enough to hinder a full range of research; the 1973 Supreme Court ruling on abortion is likely to further reduce opposition. The same can be said, but with diminished strength, about genetic engineering. There is no strong public or professional demand for research in genetic engineering, and some strong critics can be found in the scientific community.[5] That means there will most likely be no large expenditure of time, scientific energy or money in or on the field, which will advance mainly as the result of the efforts of a comparatively few scattered individuals.

All of these assumptions can be summarized in one generalization: the technology necessary to develop genetic counseling and engineering is being welcomed by the public, with comparatively few of the reservations one can find associated with the development of other (nonmedical) technologies. Given this climate, it is almost inconceivable at present that Congress, say, would vote not to provide funds for further development, as it did in the case of the supersonic transport. Genetic disease is at best a burden and at worst a misery; short of the most overwhelming objections (not visible at present) there is no reason to suppose that the full power of technology will not be brought to bear against it.

At present there are approximately 1,900 known genetic diseases not counting a very large number of chromosome abnormalities.[6] Between 5 and 6 percent of all children are born with one or another genetic defect, ranging in effect from mild to lethal, and in origin from a single chromosome or gene to complex clusters of genes. Each one of these children will have to live with some degree of handicap. The parents of each of these children will have to bear burdens which other parents do not. One way or another, society will have to bear some cost for the care of these children. At present, it is possible to detect with some reliability the adult heterozygote carriers of some eighty genetic diseases;[7] it is not thought impossible that another thirty or so more will soon be detectable. Through amniocentesis, at least sixty genetic diseases are detectable *in utero*, as well as virtually all of the manifold chromosomal abnormalities;[8] another ten to twenty are thought possible of detection. High-risk populations for certain recessive genetic diseases have been identified for at least ten conditions. In light of the fact that there are some eight hundred known recessive genetic diseases, it is evident that the proportion of those diseases which can be detected is tiny. Even more striking is the fact that there are effective therapies available for perhaps only five or six major genetic diseases, and palliative therapies available for only nine others. Self-evidently, medical science is a long way from control of genetic diseases; the surface has hardly been scratched. Put another way, the burden and pain of genetic disease continues; progress to date against it has been miniscule.

If these facts about genetic disease are granted—together with the psychological probability that few would be willing, given a chance, to live with genetic disease—it might then be asked why there could or should be any question about the "control" of genetic disease. There are two answers possible to this question. The first is that, given a solid human tradition (not just the tradition of some cultures) of using technology for preservative and curative purposes, there is no reason why this tradition should not be extended to the cure and control of genetic disease; it can cripple and kill like other diseases. The second answer, which looks to a broader frame of reference, is more hesitant, and begins by asking a number of questions. What kind of social pathology might be introduced by a vigorous attempt to control genetic quality? What would the biological consequences of such an attempt be? What would be the long-term ethical and political consequences? These are all painful questions to be forced to ask, questions which can seem capricious, precious or even cruel in the face of the evident suffering brought on by genetic disease. Yet if there is to be a science of limits for technology they must be faced. There has yet to be a technology developed (including nuclear weapons) which cannot find apologists; or, with less bias, cannot be said to produce some lesser degree of suffering in the short-run. The atomic bomb was dropped on Japan—so the official justification went—in order to shorten the war and thus to save lives. There can, in short, be no science of technological limits if, in each case, the positive reasons for going forward are allowed to have the floor to themselves. The whole point of a science of limits is to engage in some systematic negative thinking, however strong the *prima facie* case for unlimited progress may appear, in order that, in the end, there can be an understanding of what can reasonably be expected from technology.

The first step in such an examination is to stand back a few steps from genetic disease and to attempt to see it in historical and cultural context. When this is done it soon becomes evident that it is a type of "disease" which has a meaning and significance that go beyond physical pain and suffering.

One needs to ask the question, exactly why are we concerned with genetic disease and how might we appropriately look upon it? I will leave to others a treatment of the various ethical problems which the detection and treatment of genetic disease pose. I want to focus my attention here on the kind of human, social and cultural perspectives brought to bear on genetic disease, and the way in which they tend to establish the framework within which the technological and ethical issues are discerned and handled. This is important not only because these perspectives are interesting in themselves, but also because ethical systems, codes and insights spring ultimately from assessments of the meaning and significance of human existence. All of us have certain images of life, fundamental stances toward reality, and sets of assumptions about the nature of things. Taken together, they help to determine the way questions are framed, priorities of interest established, emotional commitments made, fears, hopes and anxieties aroused. What image of human existence is pertinent as a framework for determining our response to the reality of genetic disease?

When I was about seven I was approached one day on a playground by a very odd-looking boy of perhaps twelve or thirteen, stumbling, misshapen and with a rather silly, though friendly, smile on his face. My reaction was a mixture of horror, fear and curiosity. I wanted to stare at him and study him, but at the same time I wanted to run home as fast as I could. I resolved the dilemma at first by moving carefully backward as he kept coming toward me, trying to keep him at a safe distance. But he kept coming and, eventually, I did run home, certain that if I did not he would "get me." He represented an alien, utterly inexplicable phenomenon.

No one of course had told me about Down's Syndrome, mongoloidism, and no one did for many years after. My parents said vaguely that he was "sick" and that "he wouldn't hurt me," but little more. That he was the child, usually hidden, of a close neighbor was something I had to deduce for myself. All the adults in the neighborhood it seems knew about him, but all tacitly engaged in a highly successful conspiracy to say nothing about the matter, indeed to evade all pointed questions asked by innocent children. Eventually, the mongoloid I had encountered simply disappeared, no doubt into some institution.

I mention this incident because it sums up one pervasive, almost instinctual human reaction to genetic disease. The defective human being has historically seemed a mysterious, awesome, terrifying creature—almost a separate, subhuman species—to be shunned altogether, explained away by means of elaborate religious mythologies, or given a special meaning, either as an omen of good or of evil. Of course we now know more about genetic disease. We have some notion now of why these hitherto strange creatures issue forth from the womb as defectives; and we think we have some ways of coping with what we are now able to classify as an abnormality.

But it would be a mistake to think that the old sense of mystery, awe and terror does not live on. Surely the pervasive parental fear of bearing a defective child is not just because of the troubles, economic and practical, of raising such a child, though that is a most serious matter. The fear seems much deeper than that, as if a defective child would represent a supreme undoing of the parents' image of themselves and reality. I am reminded of the conviction of many philosophers of antiquity that there is an irrational, irreducible "surd" element in the universe, which constantly breaks through our visions and structures of ordered rationality. Genetic disease, long before it was labeled as that, was a case in point. One might in the instance of a hunting accident be able to discern a causal logic; the accident, however unacceptable, at least made sense. Genetic disease made no sense whatever. A defective child just arrived, out of some primeval darkness. A terror before that darkness still endures.

Yet unlike our ancestors we are no longer willing to be cowed. We are beginning to fight back, and the steps being taken follow a familiar script. We try to understand the mystery, by turning it into a problem rather than a mystery; then, with a cool-headed, rational explanation in hand we set ourselves to doing something about it—we label it a "disease" and then we try to "conquer" it. The change in our image of reality is evident here. Our ancestors had no choice but to be terrorized, and thus in compensation to find some symbolic means by which the philosophical absurdity of the defective's existence could be understood and lived with. For our part, we act—fast, hard and with all the technical means at our disposal. The results are becoming evident. We are beginning to understand genetic disease, to know how to diagnose its presence, and, in some cases, how to cure it. Increasingly, it is becoming possible for those who do not want to bear defective children not to do so. Since no one wants to bear such a child, just as they do not want to contract cancer or coronary disease, we are all that much better off.

Yet at the level of broad philosophical perspective I believe we should anticipate some problems. The conquest of genetic disease proceeds from the optimistic assumption that physical reality can be understood and its aberrations minimized or eliminated. Nature can be brought to heel. That has been the assumption of the medical sciences generally, and its successes have served as the best possible arguments for some against those who still believe there is a surd element in reality. Unfortunately, though, the fact of the matter is that we have—as indicated by the statistics given earlier—just begun the "conquest" of genetic disease; it may be centuries or millennia before the last defective child is born into the world. If a heuristic optimism is a necessity in order that the struggle against the disease go forward, this ought not to blind us to the fact that there is far more of what we do not know than of what we do, and far more genetic diseases that we have not conquered than we have.

In the meantime, during a great transition period which has no discernible end point, what ought to be the philosophical significance we attribute to genetic disease and genetically defective human beings? What ought our image of the world to be and what place ought genetic disease to have within it? Let me point to some items on the present cultural horizon which seem to me to bespeak some dilemmas which need to be considered.

First, we still speak much of freedom of parental choice in the use of genetic information, and it is still common for genetic counselors to reassure parents that a defective fetus or child is "not their fault." Yet it is possible to detect tendencies which could eventually rob people of their choice and "blame" them for the defective children they bring into the world. If there is something of a paradox here at first sight, it soon disappears. For the corollary to giving people freedom of choice is to make them responsible for the choices they make. It is then only a very short step for social mores to begin distinguishing between "responsible" and "irresponsible choices"; overt and covert pressures concerning the "proper" or "ethical" use of freedom begin to put in an appearance.

Thus while in principle the parents of a fetus with a detected case of Down's Syndrome are still left free to decide whether to carry it to term, it is not difficult to discern an undercurrent in counseling literature and discussion that would classify such a decision as irresponsible.[9] This is amplified in a subtle way: amniocentesis and subsequent abortion are said to be "medically indicated" in such cases, as if what is essentially an ethical decision has now become nothing but a medical one.[10]

I am not concerned here with whether the decision in this or similar circumstances *is* irresponsible, or whether an abortion ought to be performed. I only want to point out that if and when we begin holding parents responsible for the children they bring into the world, and then as a consequence blaming them for what we take to be "irresponsible" choices (however free), we may then have a result worse than that which flowed from the superstition of our ancestors. We will then share with our ancestors the view that a defective child is a curse but then, unlike them, provide no comfort whatever other than the ascetic reward of praise or blame for socially acceptable behavior in the face of the curse. In a word, it is simply not clear to me that in this circumstance an image of the world which, based on fierce rationality, would hold every person responsible for every jot and title of his or her behavior, is altogether preferable to one which, while it could give them no choice, did not judge them either. If society begins blaming and thereby socially coercing those who make what we term "irresponsible" choices, then the freedom of choice now extolled in genetic counseling would become a mockery.

Second, while it has no doubt always been the case that parents of defective children have had to count the costs, the introduction of modern cost-accounting and cost-benefit analysis into the genetic equation adds a distinctively different element. We can now, quite literally, put a price on everyone's head, working out the long-term financial costs to individuals and societies of caring for a defective child. But observe a curiosity. It was counted a great advance of the modern mind when a bookkeeping God, with his minutely maintained ledger of good and bad deeds, was noisily rejected. Yet here we are, beginning to keep our own books, and using them increasingly as a determinant in deciding whether or not defectives should be allowed the privilege of birth, and their parents the privilege of parenthood. Moreover, we seem to have forgotten the reason why the bookkeeping God was rejected—because it seemed eminently unjust, insensitive and outrageous that a score card be kept on human lives. Indeed, we are even worse than that old God; for at least in his ledger everything was supposedly recorded. But our cost-benefit analysis totes up only one item—what the financial liability of the defective will be, what he will cost us in terms of taxes, institutional facilities, the time of medical personnel, and so on.[11]

Needless to say, this kind of reckoning is prone to be weak in comparing the costs of the defective against the cost of other patently foolish public and private expenditures which are accepted with barely a word of protest. What is society spending on cosmetics this year? Nor does this kind of reckoning have much to say about ways of reconstructing the social and political order to make possible a more humane treatment of those considered a social liability—the poor, the aged, the unfit. No, for the kind of cost-benefit analysis which seems to be emerging in genetic calculations goes only in the cost direction. It is seemingly assumed that the benefits to a society which decided simply to bear the costs of humane care are either nonexistent or simply too intangible to be worth much bother.[12]

Third, behind the human horror at genetic defectiveness lurks, one must suppose, an image of the improved human being. The very language of "defect," "abnormality," "disease" and "risk" presupposes such an image, a kind of prototype of perfection. In the past there seemed little which could be done to realize the prototype; it was contemplated as a Platonic form, well out of reach. Eventually, a kind of wisdom developed. Since it appeared clear that even if there is a perfect human being to be conceived or imagined, the price of pursuing and imposing it is very high in terms of the ways it can change our response to actually existing human beings. For these reasons, the eugenics movement of the late eighteenth and early nineteenth centuries was generally rejected. Human diversity began to be appreciated and justified. In fact, behind modern society's at least verbal commitment to civil rights, to justice for all, lies a rejection of images of individual perfection and uniformity. A similar pattern has been manifest in even the purely scientific sphere, with a rejection in biology and zoology of monotypical modes of analysis and classification (which stressed a fixed, static concept of type) in favor of a populationist approach (which makes room for a richer appreciation of individual variation).[13]

Yet because of the advances in the detection and cure of genetic disease, the ghost of the perfect human being, once sensibly laid to rest, is putting in his appearance again. We are beginning to see evidence of what this means: a heightened anxiety on the part of prospective parents about bearing a defective child; increased social pressure against those who would bear a child deviating from the norm of perfection; and, at what I hope is not the leading edge but a fringe only, hints that the sensible society of the future will of course deny parents the right to bring into the world any child who could not measure up, who would be a burden upon society.

There have surely been abuses and a good deal of just plain silliness in recent rejections of purported ethical absolutes and ethical objectivity, of fixed codes, of rigid notions of the "human" and the "nonhuman," of the worthy and the unworthy, of static notions of perfection and imperfection.

But in many respects this rejection has helped to soften life, to provide a defense against the totalitarian reformer, to widen our appreciation of the value of human differences, variations and dissimilarities, to lead us to construct visions of political and social orders which can encompass and profit from variety. An image of man based on a pluralistic rather than a monistic conception of reality has been hard-won. It would be a supreme irony if, in the name of even greater progress, there was reintroduced the old monistic, monotypical kind of thinking, this time, as before, in the name of value, good order and a mythical notion of perfection.

Fourth, every image of the world and reality carries with it a correlative view of the nature of human community. It is not for nothing that political reformers and revolutionaries (think only of Karl Marx) have felt that the condition for a change in ethical and political thinking is, first of all, a change in metaphysical thinking. An egalitarian social structure can, only with difficulty and patchwork logic, be built upon a metaphysic of fixed hierarchical realities. Participatory democracy has nothing going for it in a society which secretly believes that some people are inherently lower than others and thus not fit to govern themselves.

To say that past human societies adapted happily and well to the reality of genetic disease and defect would be a serious mistake. Yet there did develop the humane response that society as a whole should carry a share of the burden of caring for the defective, that the parents or family should not be left with sole responsibility for the care and survival of the defective. Moreover, in laws against infanticide, child neglect and abuse, there was reflected a belief that society has an obligation toward the welfare of children, an obligation which holds good regardless of their social, physical or mental condition. We can see in these historical developments a concept of community and society which tried to respond to human diversity, even that diversity represented by the grossly abnormal. So, too, it was perceived that any effective recognition of diversity—which means a recognition of actual inequalities of individual assets and liabilities in society—required a joint sharing of responsibility by all for all, so that those least equipped by nature or nurture to function would not be inevitable losers.

This cultural development represented a great milestone in human history, even if our society and most others have hardly succeeded in living up to the obligations which it entailed. Public institutions for defectives were created, only too often to treat them like animals once we gave them the benefit of admission. Nonetheless, the introduction of the very notion of common and public responsibility represented a triumph of great magnitude. A triumph, I am sorry to say, which would all be for nothing if we accept the idea that defectives and their parents have no right to burden the rest of us with their troubles, or that it is naïve to find a social solution for a problem which might, but only might, be done away with by a scientific solution.

I would not want to be misunderstood here. Nothing I have said should be construed as an objection to the further development and refinement of genetic knowledge and the art of genetic counseling. On the contrary, the suffering brought on by genetic diseases warrants nothing less than a full-scale social, scientific and economic effort to eliminate, cure or alleviate them. My concern, rather, is with the spirit in which such an effort is undertaken, with the kind of philosophical perspective which should lie behind it, and with the social context in which it is carried out.

Genetic disease will not be done away with overnight, if ever. At the simplest level, there is at least a requirement that people's hopes not be raised too high too precipitately; that will just set the stage for even more suffering. More important, though, is the question of how we are to continue living in the company of genetic disease, even as vigorous steps are taken to minimize its impact. The dilemma, which is at once psychological, ethical and sociological, comes to this: How can we manage both to live humanely with genetic disease and yet to conquer it at the same time? Both goals seem imperative and yet the logic of each is different. We cure disease by ceasing to romanticize it, by gathering our powers to attack it, by making it an enemy to be conquered. We learn to live with a disease, however, in a very different way: by trying to accept and cherish those who manifest it, by shaping social structures and institutions which will soften the individual suffering brought on by the disease, by refusing to make the bearer of the disease our economic, social or political enemy.

Our communal task is to find a way of combining both logics. That will not be easy, if only because most people find it easier to cope with one idea than with two at the same time. It will mean, for instance, simultaneously working to improve the societal treatment and respect accorded those born with defects and working to extend our genetic knowledge and applying it to genetic counseling. It will mean taking the idea of free choice seriously, allowing parents to make their own choice without penalizing them socially for the choices they make, or condemning them for those choices which will increase the financial costs to society. It will mean some care in the way we use language. As physical organisms we stand "at risk" of genetic disease, as do our children and our descendants. As cultural beings, however, we stand "at risk" also in another sense, of seeing our values, our better instincts and our humanity diseased and crippled. If odds are to be calculated, then let both sets of risks be included in our general equation. Moreover, if the "risk" of genetic disease is to be calculated, and better calculated, let it be remembered that no set of mathematical statistics can, by itself, "indicate" what we ought to do in response to the realities the statistics delineate. That requires the introduction of ethical premises, and those premises, in turn, require some coherent notion of what the human good is.

When we speak of the good of "the patient," let it not be the case, in the first place, that we use the term "patient" only for those we think desirable to treat (using some clinical term for the rest—fetus, conceptus, neonate); or, in the second place, that the decision whether or not to treat a patient not be made a matter of how we, as individuals or community, happen to feel about the intrinsic worth of that patient. Beyond that, particular care is needed when we speak of someone's or some group's "good." It is not all that easy to know what is good for the individual, for the family and for society, though one might gather otherwise in light of the many confident statements made on the subject. As a descriptive term indicating deviation from a statistical mean, the word "abnormal" can be very helpful; as a philosophical term, indicating some mandatory level of social and genetic fitness—a prototype of human perfection—it can be very hazardous.

To conquer a disease is to reflect a view of the world. It is also to create a partially new world and a new view of human possibilities. How we go about dealing with genetic disease—the kinds of counseling techniques developed, the professional consensuses which emerge, the attitudes developed toward carriers of defects and toward the children many of them will bear, the kinds of choices which emerge and the positions taken on the nature of those choices—will both reflect one world and bring another into being. That is a heavy burden to bear and we had better be aware of it.

Culturally, we live in an age which has still to leap one way or another. On the one hand, the individualism bequeathed us by the enlightenment, laissez-faire economics, existentialism and pop psychology (to mention only a few sources) has never been so strong. Each person, this individualism asserts, should be left free to act as he sees fit, with minimal interference from the state. Families should be free to choose the number of children they want and the quality and kind they want. On the other hand, the drive for community, for corporate responsibility, for mutual co-interdependence has rarely been so strong either. Each person, this communitarian drive asserts, should recognize that he lives within a tight web of mutual responsibility and obligation; neither rugged individualism nor single-minded crusades for personal self-realization can lead to a genuine human community. Families should have no more children than an overcrowded world can tolerate, nor should they bring into the world defective children who will be a burden on society.

That these two cultural movements are now engaged in a great struggle can hardly be doubted. But the eventual resolution of the struggle is far from clear. The daily battle reports bring some confusing results. On the economic front, old-fashioned dog-eat-dog capitalism is being put to rout by the welfare state. On the civil liberties front, there seems to be a stalemate. In the prisons, iron-fisted law and order is being imposed by gun and club, backed up by crusades against crime in the streets. But in the physician's office, the taboos against contraception, sterilization and abortion are now disappearing in the name of individual liberty. Yet on the ecology front, the old individual liberties are under sharp assault. The once-granted right to dig, cut, dam, trash and pollute are no longer dispensed with abandon.

These confusing struggles provide the cultural backdrop to the problem which concerns me here. Whatever genetic solutions our society achieves, good or bad, will inevitably reflect and be a part of the larger ethos of the society, its attitudes, its values and, in particular, the way it resolves the age-old struggle between the individual and the community. Let me try to translate this struggle in the context of two key questions: Who should be born? and Is procreation a right? On occasion, these two questions are joined, as if they amounted to the same thing. They should not be confused. "Who should be born?" raises different issues from the question "Is procreation a right?"

To ask, *Who should be born?* is to ask who has a right to be born, whom do we want to admit to the human community, and what conditions of health and normalcy need to be met to merit the title of human. Here the emphasis can go in two directions. First, it can go in favor of the person who is to be born, putting his or her rights first, demanding in turn that the community not set aside those rights in the name of its own societal needs. This was the thinking, broadly taken, which lay behind antiabortion statutes, and which would lie behind resistance to a eugenic use of new developments in the detection of genetic disease—aiming to eliminate the "unfit." A variant of this emphasis might well allow abortion under many circumstances, or even a mild foray into eugenics, so long as one condition could be met. That condition would be the giving of a primacy to conflicts of individual rights. Thus abortion would be morally allowable if it could be shown that the individual rights of the fetus were in conflict with the individual rights of the mother. In short, the only conflict to be admitted would be a conflict between or among individual rights and liberties.

A second direction is, however, possible and increasingly apparent. This places the emphasis upon the needs of the community and of society, rather than on individual liberties. Take the case of abortion for a fetal defect. Not long ago the main argument turned on whether it was the right of the mother, or of the parents, to choose abortion in the case of fetal defect in order to protect their own welfare and that of their existing family. Now the emphasis is beginning to fall more heavily on whether the welfare of society as a whole is enhanced by abortion for fetal defects. Thus there are an increasing number who would argue that even if an individual couple is willing to run the risk of bringing a defective child into the world, and to bear the psychological burden of caring for it, it would nonetheless be antisocial for them to do so. They would be imposing upon society all of the high costs which go with the care and raising of defectives. In short, the whole debate shifts from the arena of competing individual liberties to that of the common good versus individual liberties.

If this is the case concerning the question *Who should be born?* it is no less so with the question *Is procreation a right?* The old argument turned on whether the right to procreate should be a matter left entirely to individual conscience, with individuals and couples set free of state intervention in their procreative decisions. We know how that argument came out, with a decisive triumph for that side which pressed for freedom of choice. The new argument, by contrast, focuses its attention on the needs of the world and of particular societies. Given an excessive population growth, and given the high costs of caring for both defective and unwanted children, an unrestricted right to procreate has been sharply challenged. What may be in the interests of individuals is not necessarily in the interests of society or the future of the species. Thus some would argue that either indirect pressure or outright coercion be exercised against those whose procreative habits pose dangers to the common good.

Tay-Sachs disease raises all of these questions, even if, unlike some more prevalent genetic diseases, it touches the lives of only a comparatively small number of people. Tay-Sachs disease occurs as the result of a deficiency of the enzyme hexosaminidase A.[14] It primarily occurs among Ashkenazi Jews, with reported incidences of 16.9 to 26 cases per 100,000 population. A child afflicted with the disease develops normally for the first six months of life. Thereafter, a rapid mental and physical deterioration sets in, with death inevitable by the fourth or fifth year of life. It is estimated that it costs approximately $25,000 a year to care for an afflicted child. Three prototype positions concerning Tay-Sachs disease can be sketched. First, it is possible to argue that even though Tay-Sachs disease is fatal at a very early age, and though the care of an afflicted child is expensive, and though the trauma of bearing such a child is severe, its right to life is inviolable. This line of argument would tell against the use of abortion to eliminate those fetuses in whom the disease has been detected by amniocentesis. It would also imply a social obligation to care for Tay-Sachs children. It would not, however, be incompatible with efforts to detect carriers, married or unmarried, and urge them not to marry, or if they do, not to have children.

Second, it is possible to argue that carriers of the Tay-Sachs gene have as much right as anyone else to marry and as much right as anyone else to have children. Thus if an afflicted fetus is discovered, it should be the right of the parents to elect an abortion and, subsequently, to procreate again in the hope of having a normal child. For those parents who decided to bear an affected child, society would have an obligation to do what it could to help them care for the child.

Third, it is possible to argue that while carriers of Tay-Sachs may have a right to marry, and may have a right to procreate, they do not have the right to bring into the world a Tay-Sachs child. Thus they would have an obligation to undergo whatever tests were necessary for the detection of carriers and of affected fetuses, and an obligation to abort any afflicted fetuses. The grounds for this position could be twofold: that it is a wrong toward the child himself to allow his birth into a life which portends an inevitable early death; or that it is wrong to inflict upon society the costs that, in the end, it will be bound to bear in caring for the child. It would, therefore, be appropriate for society to exert severe pressure against irresponsible parents or to coerce them directly.

Variants on these three positions are of course conceivable, but I will leave them aside. It is more important, I believe, to see if we can isolate the range of consistent and reasonably integrated positions which might be taken toward the disease. For these purposes, we can also leave aside the fact that the disease most heavily affects Ashkenazi Jews, and that within the Jewish community abortion is on the whole seen as a morally acceptable option in the case of detected disease in the fetus. Other genetic diseases, which affect much larger and more diverse communities, will raise more difficult social issues.

If we grant that there is an unlimited right of individuals to procreate, both in terms of number and in terms of the genetic quality of children born, then there can be no posing of a conflict between individual and society; the conflict is precluded from the outset by positing an absolute primacy to individual desire and choice. Yet it might well be argued that, while there exists a right to procreate, this right is by no means unlimited; it should not be posited as an absolute, inviolable right. This position would be compatible with allowing everyone to have some children (e.g., two) but not an unlimited number of children.[15] It would also be compatible with allowing everyone to have a child which would not pose an uncommon burden on society, but not allowing them to have a child which would. Yet it is far easier to make the moral case for a limited right concerning the number of children individual couples could have, than it is to make the same case concerning the genetic quality of the children they might have a right to.

The primary difference in the two cases is this. In the

instance of excessive population growth, each additional child beyond the socially determined optimal average family size imposes an equal burden upon society; thus no one of the extra children as such poses a greater or lesser burden (I leave aside here the question of whether a child born into an affluent technological society creates a greater burden on the world than one born into a poor nontechnological society). A social decision—"mutual coercion mutually agreed upon"—to limit family size, while it may result in some discrimination against some *parents,* does not result or spring from a special hostility toward or discrimination among any given potential children. It is not the child as such which is unwanted, but the child (or better, the class of children) which imposes intolerable burdens on the rest of the society or world simply by existing. The excessive child is, as an individual child, no better or worse than any other child, and the burden he is thought to impose on society is not the result of any special characteristics of that child.

The matter is otherwise in the case of a desire to restrict the right of parents of potentially defective children to procreate. In that case, it is a *particular* child which is not wanted, one who by virtue of having certain characteristics (low I.Q., physical abnormalities, care requirements) is thought to impose "burdens" on society. At that point, values are assigned to different kinds of lives, based upon certain characteristics thought undesirable and upon considerations of the *kinds* of lives society should or should not be willing to tolerate and support. Population control, therefore, even of a coerced kind raises fewer ethical problems, and has less ominous implications, than control of genetic quality. When uncontaminated by genetic considerations, population control says no more than this: any kind or type of child, with any characteristics, can be born, so long as the average family size into which that child is born does not exceed two. Genetic control, by contrast, says this: no matter what the other circumstances (demographic, sociological, etc.), no child who is defective (or seriously defective) should be born—because that kind of child will pose a burden on society. It would be an insult to the reader to think his or her imagination and sensitivity are so deficient as to be unable to visualize the kind of social pathology which would be introduced into society by government regulations concerning the genetic quality of children couples would be allowed to bear. A social experiment of a similarly eugenic kind was carried out in recent history in Nazi Germany; the results did not commend themselves to most human beings.

One major flaw in most discussions of the larger problems of who should be born and the right to procreate is that quite disparate elements are often lumped together, as if they all raised the same kinds of issues. The first step necessary in clarifying the issues is to sort them out. The most important distinction concerns the different participants in the problem. There is the potential child himself, the family of the child, and the larger society. A failure to distinguish among the rights and obligations of each is the source of considerable confusion. It is not enough to say, in the usual rather loose way, that a Tay-Sachs child (or a mongoloid) should not be born (*a*) because the child will suffer, (*b*) because he will cause the parents to suffer and that (*c*) the financial and other costs to society of caring for such a child are terribly high.

The reasons for not talking in this loose, all-inclusive fashion should be evident. If an affected person has a right to be born and to live, then this right cannot be set aside simply on the grounds that the child will cause the parents to suffer; it has not been part of our tradition to deprive others of life because of the burdens they impose on those around them. Moreover, it has increasingly been thought the function of government to protect lives and, through use of the power of taxation, to raise such funds as may be necessary to support those whose lives are disadvantaged.

From quite another perspective, even if one grants parents the right to decide whether or not to abort an affected child, this right would become all but meaningless if there did not exist the option of care for a child not aborted. The existence of societal support for the parents of defective children is necessary if they are to have any freedom of choice at all. Finally, if we speak of the "interests" of society, they are varied. There is the interest of allowing people freedom of choice, the interest of protecting the welfare of the present and future generations, the interest of avoiding the introduction into society of a predication of rights based solely on cost-benefit calculations, or the introduction of social pathologies. I do not mean to deny that there can be genuine conflicts of rights and interests. The right to life of a fetus can conflict with the rights of parents and their children to their own life, either physically or psychologically. The rights of society to its own preservation can conflict with the rights of fetuses and parents. If individual and social decisions are to be made, they can be made only by taking care to specify exactly what rights and interests are at stake and how.

When we ask, "Who should be born?" what exactly are we asking? Are we asking whether there is some ideal child who should be born, one without spot or blemish, aberration or defect? If so, one must then ask whose standards of normalcy and perfection should be followed? Beyond that, we must also ask just what the very idea of "perfect" or "normal" children means. Does it mean that only their lives are worthy of respect, that to be normal signifies not just health but some normative value-laden state of affairs, whereby the worthy and the unworthy are to be sorted out? If so, is this the kind of society we want to live in? How far is the idea to be pressed—to the point of denying those who would bring a defective child into the world the right to do so, thus predicating any right to procreate on the quality of the child likely to issue from that procreation? How far are we to go in adding up the costs of the care of defective children? Are we to do the adding independently of the costs of everything else we do in society? If it costs $25,000 a year to care for a Tay-Sachs child, is this really very "expensive" in an affluent society, or is it expensive only because our expenditure on other dubious ventures—excessive defense budgets, luxury items and the like—are so prodigious? What does it mean to say, in an affluent society, that some costs (but not others) cannot be afforded?

These are only questions, and my own answers are not terribly important. What strikes me, however, is that these questions are so rarely asked. Perhaps in the case of Tay-Sachs there are credible answers. But when I hear discussions of other diseases, say Down's Syndrome, I am not encouraged by the kinds of answers our society is increasingly prone to serve up. I am told, for instance, that we owe it to the fetus to abort it if it has Down's Syndrome. Yet when I read of the actualities of Down's Syndrome it becomes clear that most mongoloids are happy, that many have a minimally adequate intelligence level, that many can be trained for simple jobs, that they are capable of giving and responding to affection.

That does not sound like a life of suffering to me. Most likely, though, it will mean a life of suffering for the parents. But, if so, that is a very different matter; a riddance of the child with Down's Syndrome serves the parents and not necessarily the child himself. That should be said and very clearly, at least so that we might know just what is really at stake. Even in the case of Tay-Sachs disease, far more severe than Down's Syndrome, the suffering of the child itself is apparently not great; the course of the disease brings a mercifully quick degeneration of cognitive and affective faculties. The greatest suffering is on the part of the parents. I myself feel that the parents would have a moral right to turn to abortion in that case and for those reasons. But I would hope that no one would be fooled into thinking we were really acting for the sake of the child. Nor that anyone would be fooled into thinking that we were doing anything other than taking the life of the fetus in order to preserve the welfare of the parents.

I have argued that the way in which the question before us will be resolved will be closely related to the way society decides in the future to weigh the rights of individuals against the rights of society. I doubt that any happy balance can be achieved; all solutions are bound to be temporary, just as they have always been in the history of mankind. But until someone has presented far stronger arguments than any now available on the overriding rights and interests of society in the case of genetic disease, I believe we can live with the present bias toward individual rights, at least in the case of allowing parents the right to bear defective children. It is difficult and painful enough to try resolving conflicts between individual rights; we have hardly learned how to do even that much. To thrust that quest aside in the name of some nebulous greater social good—the money we could save if we did not as a society have to support defectives, as a most notorious example—would bring few benefits and much harm.

Those last points bear on two final distinctions I want to make. First, I have argued that parents should have the right to bear defective children, even if the social costs to society of allowing them this freedom is high. A correlative right of parents is that of not being *forced* to bear defective children. This means allowing them the option of genetic counseling and subsequent avoidance of procreation if a genetically defective child would be likely, and the right to abortion if a defective child is prenatally detected. That argument alone is sufficient to justify further genetic research, and an extension of genetic counseling. The ultimate goal is to allow parents the chance to avoid having defective children, while also giving them the chance to have those children if they knowingly so choose.

Yet—and this is my second distinction—the asserting of that kind of freedom by no means entails an unlimited extension of freedom to choose genetic quality. In particular, the right to choose whether or not to bear a *defective* child is not the same as a right to choose to have an optimal child, the kind of child which might become possible should genetic engineering make rapid advances in the decades ahead. For there is no way to tell what an "optimal" child would be, either in terms of individual welfare or in terms of the welfare of the gene pool. To allow the individual desires of parents to fuel the engine of genetic engineering would be to start with no norms whatever, to open the way to the greatest possible capriciousness. If a case is to be made for genetic engineering—and someone other than I will have to make it—it can only be a case based on the good of the community as a whole, and beyond that on the good of the species. The primary arguments in favor of genetic engineering have turned on what some individuals would like (e.g., cloning). Nothing seems to me more hazardous than ventures to improve the human lot by shortcut methods, and to base these methods on individual idealized notions of perfection and optimization; that has been the primary failing of the technological mentality, where the tyranny of unrestrained individualism has left the rest of the world to deal with the results.

## NOTES

1. Albert Dorfman (ed.), *Antenatal Diagnosis* (Chicago: University of Chicago Press, 1972); Maureen Harris (ed.), *Early Diagnosis of Human Genetic Defects: Scientific and Ethical*

*Considerations* (Fogarty International Center Proceedings 6; Washington, D.C.: U.S. Government Printing Office, 1972).

2. See Marc Lappé *et al.*, "Ethical and Social Issues in Screening for Genetic Disease," *New England Journal of Medicine,* Vol. 286 (May 25, 1972), pp. 1129–32; N. E. Morton, "Population Genetics and Disease Control," *Social Biology,* Vol. 18 (1971), pp. 243–51; James R. Sorenson, *Social Aspects of Applied Human Genetics* (New York: Russell Sage Foundation, 1971).
3. See Leon R. Kass, "The New Biology: What Price Relieving Man's Estate?" *Science,* Vol. 174 (November 19, 1971), pp. 779–88; Joseph Fletcher, "Ethical Aspects of Genetic Controls," *New England Journal of Medicine,* Vol. 285 (1971), pp. 776–83; Paul Ramsey, *Fabricated Man* (New Haven: Yale University Press, 1970).
4. See Bruce Hilton *et al.* (ed.), *Ethical Issues in Human Genetics: Genetic Counseling and the Use of Genetic Knowledge* (New York: Plenum Press, 1973).
5. *Cf.* Richard Roblin and Theodore Friedmann, "Gene Therapy for Human Genetic Disease," *Science,* Vol. 175 (March 3, 1972), pp. 949–55; James D. Watson, "Moving toward Clonal Man," *Atlantic Monthly* (May 1971), pp. 50–53; Martin P. Golding, "Ethical Issues in Biological Engineering," *UCLA Law Review,* Vol. 15 (February 1968), pp. 443–79; Marc Lappé, "Moral Obligations and the Fallacies of 'Genetic Control,'" *Theological Studies,* Vol. 33 (September 1972), pp. 411–27.
6. Victor A. McCusick, *Mendelian Inheritance in Man* (Baltimore: Johns Hopkins Press, 1971), p. ix.
7. Max Levitan and Ashley Montagu, *Textbook of Human Genetics* (New York: Oxford University Press, 1971), pp. 744–46.
8. Dorfman, *op. cit.,* p. 23.
9. See, for instance, the statement of Dr. Bentley Glass in his 1971 presidential address to the American Association for the Advancement of Science: "No parents will in that future time have a right to burden society with a malformed or mentally incompetent child" (*Science,* Vol. 171 [1971], p. 628).
10. See Herbert A. Lubs and Marie-Louise E. Lubs, "Indications for Amniocentesis," in Dorfman, *op. cit.,* pp. 2–27.
11. See, for instance, the following calculation: ". . . the field which

appears to offer the greatest benefits from antenatal diagnosis is cytogenetics, and certain calculations of cost are appropriate. In 1967, 300,000 infants were born to mothers thirty-five and older—10% of all infants. At a cost of $50 per infant for amniocentesis and preparing two karyotypes, it would cost $15,000,000 to survey this group of pregnancies. If 1.5% were abnormal, 4,500 chromosomally abnormal infants were born to these mothers. If the costs of special medical care and education may be conservatively estimated at $2,000 a year, these special costs over twenty years will be $180,000,000. Can we afford not to make an investment that is this productive?" (Lubs and Lubs, in Dorfman, *op. cit.*, p. 27.)

12. Most cost-benefit analysis in the field of medical genetics is crude, theoretically deficient because of a failure to compare the costs of two or more alternative responses to the same problem, a *sine qua non* of adequate cost-benefit analysis.
13. Cf. Ernst Mayr, "Species Concepts and Definitions," in Ernst Mayr (ed.), *The Species Problem* (Washington, D.C.: American Association for the Advancement of Science, Publication No. 50, 1957).
14. "Tay-Sachs Disease: Prenatal Diagnosis," in Dorfman, *op. cit.*, pp. 175–84; and Michael M. Kaback and R. S. Zieger, "The John F. Kennedy Institute Tay-Sachs Program: Practical and Ethical Issues in an Adult Genetic Screening Program," in B. Hilton *et al.*, *op. cit.*
15. This distinction is developed by Arthur J. Dyck, "Population Policies and Ethical Acceptability," in Daniel Callahan (ed.), *The American Population Debate* (New York: Doubleday, 1971), pp. 363–64.

# CHAPTER 9

## *Prohibitions and Interdictions:*
### *TOWARD TECHNOLOGICAL LIMITS*

THE NATURE OF THE ETHIC I want to propose here is not the kind that can easily be developed into a system or even fully articulated; it is meant to make us feel guilty and to hesitate. That ethic would affect the cultural super-ego more than anything else, and even if it could partially be defended by rational justification, would be useless if it had to stand trial at every decision-making juncture. If nothing else, we should by now know that one critical source of our present crisis of values is the demand that every action be justified to its metaphysical roots, every decision subjected to open discussion of the acceptability of every possible option, every possible value analyzed both for its validity and for its meaning, every motive examined for its rectitude, cultural bias and psychological satisfactoriness. The result of this pressure on all forms of ethical thinking, on all moral rules and inclinations, on all systems for understanding reality, is not better ethical thinking but worse. It leads either to outbursts of emotive fervor: for the pressures of that much thinking and examination cannot be borne—"gut" takes over and is easily legitimated; or to paralysis of all action—for no purported ethical action can survive the acid of that much examination in a culture where each person is forced to create his own world of meaning and value, to compete against the criticism of every other person forced to do the same. We are left with the vacuum of honesty and authenticity.

To be sure, contemporary systems of ethical thinking need to be examined and reexamined, together with the explicit value theories and commitments which underlie the beliefs and behavior of individuals and groups. We can hardly afford to give up our rationality, particularly that very hard and tough rationality which is restless with the usual enthusiasms, explanations, theories and evasions. Yet we cannot depend upon it to get us through; its record, while not terrible is by no means splendid. Most needed is a re-creation of the cultural super-ego in its response to technology, which could in its roots, training, shaping and implications create the social possibility of well-tutored emotions, healthy repugnances, benign inclinations, intuitive realism, sensible optimism, creative sublimations, suitable self-limitations and self-censorship, and an internalized wisdom strong enough to be proof against evanescent cultural guruism and messianic self-liberation.

That is an impossible order, particularly since no one, including myself, has any clear idea at all how one goes about reshaping a culture (far harder than reshaping a political or economic order), much less reshaping a cultural super-ego. There are hazards in even trying, witness the rise of mystical Fascism in the 20s and 30s and the current tyrannies of a commercially exploited and now captured Con III counterculture, both of which purported to establish a new set of values. But since the hazards of continuing on our current course, where group ideologies and individual fanaticisms rob society of any possibility of creating a common culture (ensuring an eventual high-minded destruction of each other), there is no real choice but to try.

The path to be taken in this effort must be essentially negative, aiming less at envisioning and creating the good than in understanding, limiting and forestalling evil. A sense of the good, however inchoate, can be adequate; it is evil which demands pinpointing. The power of Freud's contribution to the question of civilization was his unwillingness to engage in easy talk of liberation, to raise the question of altruism and its widening drive bonds for community in the most delicate and tentative way only, reserving his most powerful examination for those deepest of human pathologies which stand in the way of happiness. That is the right balance between realism and vision in technological societies. This is only to reiterate the need for a "science of limits," which seeks less for what should be done than for those boundaries which should not be transgressed in the process. While one could no doubt work with the concept of limits in every sphere of contemporary human activity and thought, I will confine my comments here to the need for creating a science of technological limits, a cultural super-ego which would sink the roots of that science deeply into the "unwitting part," and a social and political order which would make such a science humanly viable and valid in day-to-day life.

There are two reasons why a science of technological limits is needed. First, in order that limits can be set to the kinds of hope and expectations which technology has historically engendered, hopes which show a pattern of constant escalation regardless of the degree of achieved technological progress, and hopes moreover which *because* of the escalation are never realized. Technology has promised (or been thought to promise) transcendence of the human condition. That is a false promise, incapable of being fulfilled whatever the future degree of technological development. Human desires are infinite, and cannot be achieved by the finite means of technology. Nothing is more important as a setting for a science of technological limits than the most stark confrontation with that reality principle. Second, a science of technological limits is necessary in order that the pathologies resulting from technology can be controlled. The environmentalists and ecologists have already made a good start in the long process of educating technological man to the direct and indirect physical harms of thoughtless attempts to dominate and exploit nature. The technology-assessment movement could be another important ingredient in that educational process.

Yet to be undertaken, however, is an analysis of the social pathologies which technology brings in its wake. They can be introduced either by attempts to impose new technologies (as in mandatory screening programs for genetic disease, or proposals to deny the right of parenthood to those likely to produce undesirable, socially burdensome children); or by attempts to correct the harms wrought by technology through the introduction of socially discriminatory and punitive sanctions against those who, not having introduced the technology in the first place, are supposed to compensate personally for its extant hazards (as in proposals to make parenthood a "privilege" rather than a right). While technology can and does cure, save and free, it can also become the vehicle for the introduction of new repressions in society, both because it provides ever more precise methods of controlling human behavior and also because fear in the face of its excesses can engender social terrorism as a defensive, corrective response.

Any possibility of a science of technological limits will be dependent upon establishing the right context for examination and a suitable perception into the conditions necessary to make limits possible. The most important perception is to understand and accept the fact that man is a technological animal. It makes no sense to talk of *Homo sapiens* without technology nor to distinguish man from technology. There can be no place for the sentimental romance of turning the human back on technology, or returning to the style and goals of earlier, nontechnological societies, of trying to "do away" with technology. Mankind would, in almost every respect, be the loser were that possible. Technology has made life better and can continue to do so. Properly understood, it is still possible to talk rationally of technological "progress," of new scientific "conquests," of still further "improvements" in the human condition because of technology.[1]

A science of technological limits cannot be built upon an antitechnological spirit, which has always turned out historically to be self-defeating, setting the stage for a new spiral of unrestricted expectations from technology. The science must be built upon a grasp of the positive benefits and powers of technology, which at the very least—and that is quite enough—enable human beings to adapt themselves better to nature, to protect themselves from its hazards, to correct its deficiencies, and to investigate its significance. Together with the cultural, political, philosophical and religious systems which the human race has developed, science and technology can take their place among the greatest of human achievements. This is only to say that a science of limits must take full account of the good technology can achieve. The harder question is to know what it *cannot* achieve, and to find those limits which will enable technological man to know when to turn back and thus when to seek the fulfillment of his hopes elsewhere.

To speak of "limits" at all is to introduce a dour note, one which should be embraced in all its rigor. I mean by a "science of limits" the following: a system of prohibitions, denials and interdictions which establishes the limits of technological aggressiveness, the limits of technological hope, and the limits of technological mandates (social and individual). The word "no" perfectly sums up what I mean by a limit, a boundary point beyond which one (we) should not go. I will not pretend here to any special wisdom on the details of that No in every (or most) given cases. The problem with both the tyranny of individualism and the tyranny of survival is that neither know how to say No in the right way. Individualism does not know how to say No to the private self and its desire to outwit and circumvent the needs of the whole community; it can only say No to society, not to itself. The tyranny of survival is that it does not know how to say No to the needs of the community and the species; it can only say No to individual needs and desires.

A science of limits must, as a minimal demand, be able to establish the legitimacy of prohibitions, repressions and interdictions, but then know also how to distribute them properly between and among individuals and communities. The greatest moral failure of the technological mind, springing from individualism, is that it does not know the meaning of No. Assuming that all things possible are desirable, and all things desired are valuable, it has set its face against restraint and limitation, asserting that the establishment of any limits would subvert its sacred mission.

What Stuart Hampshire has said about modern utilitarianism, underlying much technology and also knowing no fixed limits, can appropriately be repeated here:

> Modern utilitarians thought that men have the possibility of indefinite improvement in their moral thinking, and that they were confined and confused by their innate endowments of moral repugnances and emotional admirations. There was a sense of the open future in all their writings. But hope of continuing improvement, if it survives at all, is now largely without evidence. Lowering the barriers of prohibition and making rational calculation of consequences the sole foundation of public policies have so far favored, and are still favoring, a new callous-

> ness in policy, a dullness of sensibility, and sometimes moral despair, at least in respect of public affairs. When the generally respected barriers of impermissible conduct are once crossed, and when no different unconditional barriers within the same area of conduct are put in their place, then the special, apparently superstitious, value attached to the preservation of human life will be questioned. This particular value will no longer be distinguished by an exceptionally solemn prohibition; rather it will be assessed on a common scale alongside other desirable things. Yet it is not clear that the taking of lives can be marked and evaluated on a common scale in which increases of pleasure and diminutions of suffering are also measured.[2]

Hampshire's complaint that utilitarianism, by abolishing the notion of prohibitions and allowing all values to be placed on a common scale, finds a parallel perception in Rieff's analysis of modern culture:

> The most compelling interdicts are those buried deep in character. Culture is no glass-bead game. From inside his culture, a man cannot say everything—indeed, he can say scarcely anything of all there is to say. The purpose of a culture is to fight off what-is-not. Death is what-is-not, the final opening of possibility that it is the purpose of culture to close. A culture that engages itself to the inter minable opening of possibility issues its own death warrant: apostles of the opening of possibility, as "Life," have adopted the transgressive rather than the interdictory style as their own. . . . Our bio-revolutionary apostles are in league with our commercial geniuses, death dealers in the sacred game, which can only be made right by virtue of constantly reconstructed limits. Because humans are so wonderfully malleable, they require constraining orders,

> repressive truths to limit the dangers of their malleability. Repression must characterize the psyche as interdicts organize a culture.[3]

Just as one can say of utilitarianism that it can mean the collapse of all morality, so one can say, with Rieff, that "Without a science of limits, the domination of science implies a cultureless society."[4]

In designing a science of limits for technology, interdictions and prohibitions are needed at the biological, the social and the psychological levels. The place to begin is with the recognition of the biological reality principle. Because man cannot live without technology, his nature requires that he make continuing use of technology; and because his circumstances and environments change, he must work constantly to shift and improve his technology. At the same time, it is historically evident that man cannot find happiness through technology, or through thinking of himself *only* as a technological animal, capable of indefinite improvement. It is the peculiar characteristic of man that he seeks everything, that no limit can be put on his fantasies and imagination. The technological fallacy, at odds with man's biological makeup, is that it is possible through technology to satisfy all those fantasies.

B. F. Skinner offers to satisfy the dream of peace and human harmony through behavioral conditioning. If only we could give up the illusion of freedom and dignity, and the prohibitions which they have carried, mankind could be set straight and led into the promised land. The drug visionaries offer the possibility of ecstasy and transcendence, if only people would put aside their innate repugnances (socially conditioned of course) and the self-restrictions which their misguided rationality has repressively forced upon them. The genetic-engineering dreamers can speak of a better, smarter, healthier, happier *Homo sapiens.* We need only excise from our consciousness the sense that there are limits to the extent to which man should experiment with man; atavism, religious superstition and a failure of nerve are sufficient to explain our feeling that somehow there is something special, something inviolable about the human substance.

What can be offered in the face of these onslaughts of infinite promise, requiring only that we set no boundaries to our technological hope and courage, to our biological *machismo?* Only human history, with its chilling record of aggression, destruction and self-betrayal, and only the fact, explored so well by Freud, that man's unrestrained attempts to make of himself a god have led to dejection and greater despair. That we refuse to learn from the human record says much about the nature of man, part of which is by no means all bad in its refusal to become fatalistic about reality, to let past failures extinguish all hope. But this refusal to learn also means that still another generation is to have inflicted upon it impossible dreams and potentially destructive visions of unlimited possibility. These dreams and visions are always inflicted on people sooner or later, inflicted because if their promise seems powerful enough then people are expected and forced to take a part in bringing the promise to fruition. The move from arguing that it would be good if populations could be screened for genetic defects, to contending that they must be screened for the protection of society, is a very small one, particularly in societies prone to turn the technological "can" into the socio-moral "must." That kind of logic is built into well-organized, highly integrated and efficient technological societies: the full possibilities of technology will be realized only if everyone joins in; if only some join, then the technology may make little or no contribution to the social good at all. What direct coercion cannot accomplish, the social imperative of technological success can.

The first interdiction required by a technological science of limits is a sharp dampening of the unchecked and uncorrected technological imagination. This dampening should take its point of departure from the biological reality principle that technology cannot provide the full measure of human happiness, and will in fact lead to misery if, in a lust for infinite possibility, the reality principle is set aside even for a moment. This is essentially a task for the cultural super-ego, which could curb the desire for technological infinity and transcendence not by rationally analyzing each and every proposal for technological "progress" put forward (e.g., by technology assessment), but by scaling down the emotional and visionary demands made upon technology in the first place—by putting in their place a sense of limits, even a sense of guilt for demanding too much. As the early attempts to judge the benefits and harms of genetic engineering have well indicated, it is literally quite impossible to resolve arguments about whether humans would be better or worse off with cloning, genetic surgery, sex selection and so on. In that kind of situation, matters simply drift, guaranteeing that the issues will not be resolved until the genetic technology has already been developed, when, that is, it may already be too late to do anything.

I do not want to be misunderstood here. I am not proposing that we cease the careful rational analysis of the possibilities, good and bad, of new technologies; far more of that kind of analysis is needed and should be done with more skill than is presently displayed. I am only proposing that we not continue to be caught in a cultural situation in which, our technological imagination out of hand, we are then forced to make judgments on matters which in a more prudent society might not have arisen at all. The only way this could be possible would be by the presence of a cultural super-ego which imbued an innate skepticism toward technological infinity, a spontaneous sobriety in the face of futuristic scenarios which would lure us into unrealistic hopes, a healthy repugnance toward those who would by the siren song of happiness and plenty lure us away from our doubts, inhibitions and initial refusals.

The second interdiction must take its rise from the social reality principle. By that principle I mean the inherent resistance that groups and populations feel toward efforts to force excessive technology upon them, or toward bearing socially imposed penalties in order to control technology. In the face of efforts to overcome that resistance, it ought to be borne in upon people that, as social communities, limits must be set to the extent to which technologies are allowed to shape and control their common behavior, and the extent to which they are expected to take advantage of technological developments in order to solve their problems and cure their ills.

The history of technology shows that it cannot for long remain a matter of choice for individuals, societies or the species. The automobile was, when horses were still available, an optional mode of transportation; that is no longer the case. The use of contraceptives was once a matter of free choice for couples; now their use has become mandatory, not only because of population pressures but also because domestic, economic and cultural life in technological societies requires family limitation. Genetic counseling is now optional, but the signs are already there that it will soon be considered irresponsible for couples not to avail themselves of it. Examples could easily be multiplied. It is difficult to think of any older technology which has not become mandatory, either by law or, more important, by social structure, custom and pressure. That insight should recommend a modest wisdom: assume, in trying to judge the benefits and harms of any new technology, that its use will if introduced soon pass from being a voluntary matter to becoming a socially enforced requirement. And assume as well that if there are some hazardous by-products from the technology, that all will be forced to pay for them, however innocent they may have been in introducing the technologies in the first place.

The second interdiction must be a prohibition of careless meddling in social orders and structures by technologists and, beyond that, a refusal to believe that the answer to the derangements of society is technological. The critical word in that formulation is the word "careless," for there is no need—nor any practical possibility either—to stop all technological attempts to deal with individual social and medical problems. By "careless" I mean an automatic assumption that every problem, regardless of its nature, must be amenable to a technological solution (because, what's technology for?), with the ensuing ritual attempt to bring its power to bear as rapidly and massively as possible. There is no reason whatever to assume that all problems can be solved by technological solutions, and no reason to assume either that it is in every case even good to seek a technological answer. It is precisely these assumptions which need to be denied, returning technology once again to the more humble role of being simply one more claimant to the role of savior. As matters now stand, nothing is allowed to compete with technology, so deeply ingrained is the tendency to give it the first crack at any serious issue.

Moreover, there is no reason to presume that technological advances must be taken advantage of once they are available. The most noxious combination is the joining of technological possibility with demands for survival. Survival obsessions wipe out choice and options—that is their power and why they are invoked. When survival and technology join hands, a technological imperative is introduced: the technology must be used. The proliferation of nuclear arms to an ever-widening circle of nations, the demand by some that genetic counseling be made mandatory for the protection of the gene pool, the insistence that the social unrest in society requires the use of sophisticated methods of behavior control, the vision of an implantable contraceptive which could be removed or neutralized only by a positive act of government to allow a couple to have a child—all of these examples indicate the way in which technology is looked to as the key to survival, not allowing the luxury of free choice.

If any single interdict is in order, it would be one which held every technological innovator liable if his innovation passed from the optional to the mandatory stage. But that, unfortunately, is a prohibition that will never exist. At most, one can hope for a situation where people instinctively know, with a knowledge buried deep in their bones, that any dream of the new freedoms technology could make possible need to be complemented by the nightmare that society will eventually insist that they must have that freedom, whether they want it or not. That is the ultimate social pathology which technologies introduce; there can be no turning back.

The third interdiction springs from the psychological reality principle. By that principle I mean the fact that human beings are not at all likely to give up their attempts to find happiness through technology. The new burst of technological fervor, already foreshadowed in the marriage of psychological technology to counterculture mysticism and communitarianism, shows that people are not disillusioned with technology. Even if they are disillusioned in part, it is of the nature of the modern technological mind that it will try to use technology to overcome that disillusionment, no other alternatives presenting themselves. Technology is nursed and nourished on that old, banal maxim "If at first you don't succeed, try, try again."

But does not the psychological principle I have sketched both contradict and subvert the interdictions and prohibitions

I have already proposed? If people will not cease to hope for happiness through technology, how can they be expected to impose upon themselves the restrictions necessary to ground their hopes in reality and thus limit them? In one way only. There must be deeply imbedded in the cultural super-ego the profound perception that violent reactions against technology carry within them the seeds of a subsequent collapse once again before the seductions of technology. That has been the historical rhythm of the matter, where the juxtaposition of the fundamental human need for technology and the harms wrought by technology present a dilemma which cannot and will not be resolved.

The third interdiction, necessary in the light of the psychological reality principle of technology, is that there must be an intuitive censorship of visions of a nontechnological society. Nineteenth-century romanticism, with its revolt against industrialism, and twentieth-century counterculturism, with its revolt against technology, represent themes which by now have been heard once too often. The only possibility of a check to an overinvestment of hope in technology is, paradoxically, the existence of checks against an overinvestment of hopes in nontechnology. Needed is a scaling down of technology and a rational control of it—not its deification and not its denial. Interdictions against an infinitizing of technological promise, impossible of fulfillment, will become feasible only when the extremes have been once and for all excluded—the extremes of hope and despair. Technology has lacked sober support, where promises are modest but real, and sober criticism, where the attack is focused on real dangers and specific alternative forms of conduct are offered.

Freud's observation that excessively high ethical demands are destructive is pertinent in this context. Recent efforts to persuade people to forswear technology have met with little heed, limited to glossy advertising campaigns against littering, musings about biodegradable trash, and legislative efforts to enact pollution controls. Already there is evidence that the ecological fervor of the late 60s and early 70s has begun to wane; the trash continues to pile up, more automobiles than ever are being built each year, more electrical gadgets put on the market. This is hardly astounding in a society which depends for its economic livelihood on technology, and is forced to compete internationally against nations no less committed to the economic profit to be found in "better things for better living." Excessive demands upon people to give up technology cannot fail to induce apathy or outright hostility.

No substitute for technology has yet been found for the primary human business of facilitating adaptation and preservation; and no one urged to give up technology can fail to miss that fact. Nor has any substitute been found for the economic strength which technological societies provide. Any worker whose job is threatened by attempts to control technology can speak knowingly about that; and if he has a historical memory, he will know the sufferings undergone by earlier, less-technological generations. The irreversibility of technology—whatever the theory of the matter, that is the historical reality—poses the most important obstacle to its management. Since people cannot go back on technology, they find themselves impelled to go forward, to avoid returning to that which lies behind them, and to seek that which technology has yet to deliver—and will not deliver. It is thus a limit to *always* going forward that is needed; when to go forward and when not to is the question.

I have not drawn a happy picture, asking for an interdictory cultural super-ego, on the one hand, and pointing to all the reasons why the creation of such a super-ego faces enormous and possibly insuperable obstacles, on the other.

But that is the situation with which we have to work and there is no gain in closing our eyes to it. Broadsides against the abuses of technology, however accurate, have accomplished no notable change nor do they promise to. Lewis Mumford, Theodore Roszak and Jacques Ellul, preachers against technology, make provocative bedtime reading, but little more than that, primarily because they fail to take account of the psychological reality principle of technology—that contemporary man cannot and will not live without technology. Failing to see that, they build their critiques on the illusion that things could be otherwise. Of course they could, but not in our world in our time—which, unfortunately, is the only world most of us will ever have a chance to live in. An ethic of technology, desperately needed, must take its start from the tangible, ineradicable fact of technology. The ethic must be modest in its demands and culturally viable in its principles. The alternative is simply talk.

The only possible hope for the development of an interdictory cultural super-ego lies in a shift from viewing technology as a vehicle for saving and satisfying the individual to one which roots its values in the needs of the community. Let it be understood, having said that, that the communal need most worthy of suspicion is that of survival, a need which can be claimed with little plausibility by technological societies. The communal needs I have in mind are the satisfaction of basic health and economic needs, the sustaining of viable educational institutions, the possibility of meaningful work, a minimally adequate national defense system (*not* to be confused with the present defense establishment, based in great part on considerations of economy-enhancement), the preservation of natural resources, clean air and water, satisfying residential and living arrangements, and a cultural life which provides physical and aesthetic enrichment. Naturally, even to suggest a list of this kind—noncontroversial enough on its face—does not take us far toward determining what a sufficient *level* of these goods would be, or how one should define such variables as "viable," "meaningful" and "minimally adequate."

The fact that technological societies work from very high base-lines on each of such variables poses the most stubborn difficulty. People have been led to believe they need more than they actually do, a thesis that can be demonstrated by visiting countries where the base-line is lower with no notable misery among the population as a result. The main work of an interdictory cultural super-ego would be to scale down the base-line, neutralizing the sense of restlessness, unrequited desire, and obsessive attempts to achieve still more happiness through technology that are the marks of advanced technological societies. This task can only be a communal undertaking, in part because individuals are not likely to lower their own expectations unless they believe others will do so also, and in part because the value changes necessary for a downward shift in demands and expectations can emerge only from a social rather than an individual ethic.

The requirement of a fresh public morality, pointed toward a curb on the use of technology to satisfy infinite private demands and toward a curb on those community demands for the imposed use of technology which lead to social pathologies, can hardly be escaped. At the very least such a public morality is necessary because failures of technology to satisfy desires for happiness have, on the whole, led only to a redoubling of efforts to improve the technology. A constant upward spiral, operating under the belief that the good life must be just over the horizon if only we keep our technological nerve, is inevitable; and no less inevitable will be those periodic phases of hostile reaction, which only set the stage for one more surge in the spiral. That spiral must be put to an end, and the only means in sight are a public morality and a social ethic which will sink deeply enough into the communal self to impel its own innate limits and inherent interdictions.

In Chapter 6, I provided a number of reasons why the present ethical individualism cannot be sustained. It cannot handle those problems where people with diverse values must work together to deal with common problems, cannot create the necessary sense of trust which must undergird any community and cannot, in particular, deal with those problems of technology where, because their implications and consequences are communal, the values by which they are judged and controlled must also be communal. The only present way of managing value conflicts is through courts and legislatures, a healthy enough situation when there still exists some sense of a common tradition and a broadly shared range of values. Now, however, courts and legislatures are increasingly forced to play the role of adjudicating among blocs of interests with nothing in common, and especially forced to do what in great part should be done by ingrained cultural values: provide a framework of moral rules and a common ethical core for humane societal relationships. In the absence of shared norms, there exist no grounds for questioning technological innovations, no standards for judging the consequences of technological change (even when we know the consequences), and no ways of restraining a technological imagination which, we can be sure, will continue to offer nirvana to unhappy, alienated individuals and societies.

The purpose of a social ethic is to establish a mediating ground between the private values and desires of individuals and the necessary demands and repressions which life in common imposes. When those private desires meet no serious obstacles in the cultural super-ego, and are expressed with an aggressive virulence, the only recourse of society is outright, externally imposed repression; force is brought to bear. Less dramatically, all struggles must be taken to court, ensuring the smoldering hostility of the losers and a new round of attempts to circumvent the powers of government to impose sanctions. This latter process is already appearing in the assessment of technology. In the absence of any social consensus on the value of particular technologies, no option is available but a court test or a legislative struggle. Not only is this method cumbersome, it can only be reactive, dealing with problems which have already become so advanced and exacerbated that even a court or legislative decision is not likely to resolve the issues adequately, much less turn back the clock. In the end, it is the culture and not the courts which should provide the needed control of uninhibited individual desires. Since it is those desires which fire the technological establishment, the needed social ethic must be a communal undertaking, seeking to establish those limits to self-expression and self-determination without which there can be neither a culture nor a society.

The most difficult task in creating a rational social ethic which can win support and dig itself into the cultural superego is that of finding the proper balance between the rights of individuals and those of the community. Societies and historical circumstances differ, and for just that reason there is no possibility of establishing the proper balance for every community at every given moment in time. Societies which have managed to keep the wolf of scarcity at bay can afford considerably more individual freedom than those which have not; the demand for survival is weaker. Yet as we see in technological societies, the individual demands made upon technology can have the effect of creating a false sense of scarcity. If every individual wants everything and believes he has a right to it as a necessary condition of his personal self-realization, then scarcity becomes psychologically endemic, with every person so affected living under a constantly perceived threat of degradation or annihilation. That his actual survival may not be at all threatened is beside the point. He perceives it as threatened because of the demands he makes upon life, and his efforts to achieve psychological security can take a turn as terrorized and aggressive as that of any group whose literal survival is at stake. There is a painfully fine line between the egoistic and predatory excesses engaged in by the affluent in their quest for individual fulfillment and those of the Ik, a tribe obsessed with survival and the constant need for war of each against all.[5]

Yet it is much easier to proclaim the need for community- rather than individual-centered thinking than to work through the ethical difficulties this stance engenders. In my chapters on population and genetics, I have pointed to the way in which, once the community (or the species) has been given moral preeminence, there can be a very rapid descent to persecution and tyranny in the name of the common good. Is there any middle ground between an unlimited individualism, which can generate terror, and an unlimited regard for the community, which can generate no less terror? Only in two ways, by establishing upper limits to what individuals can demand and expect, and internal restraints on the harm a community may do to individuals in the name of its own welfare.

Pretechnological societies, with no scientific solace to offer, had no problem of setting limits to individual desire: there was relatively little hope that individuals could get what they wanted, and no means available for even trying. Technological affluence managed to break that cycle, allowing each individual to think that everything is possible. But when everyone begins to think that everything is possible (and deserved) for him, then the only defense of society is to think that every kind of repression may become necessary to save itself from fragmenting chaos. An unmistakable feature of our own culture is the simultaneous rise of a wider and wider range of individual liberties together with an increasingly strong, and implemented, call for more harsh methods of law and order. If the affluent middle class now enjoys many of the same kinds of moral freedom and privileges of self-determination once enjoyed only by the very rich, there still remains a class at the bottom, expected to toe the line of the values which their betters have established as socially normative.

When public morality is nothing other than the law laid down by government, and private morality nothing other than each person's musings on the good life, then community is dead, altruism nothing more than a reckless gesture, and technology king. In the absence of community, technology must be king, for there is no other in sight who offers a way out of the desert. Government will be forced to make use of technology to control society, a society clever at evading the needs of others, thus setting the stage for social pathologies. Individuals will no less be forced to make use of technology, because they must find ways of circumventing the demands of government (which in turn is trying to evade their demands), dodging the requirements of the community and the species, and finding private solace amidst public misery. The only viable public morality is one which relies on well-established general moral principles, binding on individuals and governments, setting limits to what individuals may ask for and do in their own behalf, and to what governments may ask for and do in behalf of the people.

The notion of prohibitions has come on bad times. Somehow it is thought morally unhealthy that ethical interdictions should exist, full of flat "no's" and "thou shalt not's." Better the flexibilities, accommodations, freedoms which an enlightened morality of situation, sensitivity and enlightened self-interest bring as gifts to repressed mankind. But it is doubtful that there is much satisfaction in that kind of ethical cotton-candy. The main consequence of that kind of looseness is an open field for the manipulators of taste, consumption and sensibility. They know, better than most, that there exists no basis for a cultural resistance to every passing fad, every popular shift in values and goals; they can be sure that people will behave much as they dictate.

Lacking an instinct for limitations and prohibitions, society has no real way to judge and control technology. It can deal with technology only by giving people what they want, not by withholding that which they should not have because of the perils to themselves and their neighbors should they get it. When intrinsic limits are denied, and every proposal for a limit subjected to tests which no prohibition could pass—for if one person is left unhappy, then the prohibition must be prohibited—then all possibility for rational communal behavior ceases. The stage is then set for force to be applied. The warnings of a Paul and Anne Ehrlich that abortion will become mandatory if voluntary control of population growth is not accepted must be seen for what they are: a reasonable prediction of what could happen if the need for survival is not heeded, and a splendid instance of the way in which those addicted to survival allow their imaginations to play. They would have us think the unthinkable—which of course they profess not to hope for—and thus join the long and less-than-distinguished list of those for whom the unthinkable becomes the very thinkable.

The movement from preservation to improvement technologies, and then on to destructive technologies, finishing up with compensatory technologies to redress the psychic imbalance and harm done in the preceding steps, presents a model of irrationality. Yet a model in which, the outcome absurd, each step has nonetheless seemed reasonable. It is easy enough to blame "reason" here, with its compulsion to find and take the next "logical" steps. If it is possible to ineffectively control behavior by poorly designed methods of behavior control, why not develop effective methods and good designs? If we are going to control behavior, then let us be sensible and efficient about it. If one genetically defective group of irresponsible procreators can infect whole generations of future human beings, let us take the rational step of stopping them, by enlightened law if necessary. Reason has its compulsions.

But reason is not the final enemy in technological escalation. Reason operating without valid ethical norms has always been a hazard: might makes right; to feel good is to be good. But reason without guides, combined with a blind technology, moves us well beyond the deficiencies of human logic. Logic can only operate with premises, and when those premises are drawn from an imagination fed on hope in technology, it is the logic of the latter which takes over, spurring us on to more and more—and giving us less and less.

The greatest need Freud saw was that of reconciling men to civilization. The individualism underlying technology has made that task more difficult than ever. Technology has given the individual arms with which (he thinks) the better to fight off civilization and its repressive demands. The danger is not that people actually have such arms—for they are transitory only and never fully effective even at their best—but that they think and act as if they do. That enables them to work continually to put civilization out of their mind, to pursue their private pleasures with undivided attention. That they are dead before they even come close to succeeding is true enough; but that is a bit late to begin acting on other insights, and hardly of much use to those of us who live on, amidst the chaos they have created. The possibility of altruism, which Freud defined as an impulse to merge with others in the community, is ever hard-pressed; egoism drives us in another direction. Indeed "altruism" is not the best word for what is needed, which is something more than a willingness of individuals to give up some of their claims for a final and total self-satisfaction. More to the point is the need to admit the valid claims of the community, which even if they are of necessity repressive can also provide their own measure of satisfaction (for there is, after all, some satisfaction in survival and even more in being at one with others, however fleetingly).

Freud saw a possible reconciliation between Eros and altruism, though he did not spell out how it would be possible. He only saw that civilization *requires* a reconcilement, however fragile, however tentative, however much under constant strain. Rieff is correct in his observation that a culture is a system of controls, even if there are remissive phases; there cannot be a culture without controls. Our society has believed that, with technology, controls could be abandoned, in fact opposed as a positive enemy of self-realization. That was a myth, highly dangerous; but it lives on, finding new adherents with each passing generation. It will have to be demolished, for with each succeeding generation the dangers of technology increase; not because technology in itself is inherently evil—it is not—but because the tyranny of technology lies in its power to lure us away from those other parts of our nature which know, as well as anything can be known, that happiness is a distant dream, and if not an illusion then at least a far-fetched idea.

If one is looking for a way to reconcile human beings to civilization, that is not a bad insight with which to work. While it may not send our spirits soaring, it may do something as valuable. It may make us see that we cannot live as isolated individuals, trapped with our boundless hopes and technological schemes; that is a sure and constantly demonstrated route to misery. It may make us see that we must live as a community, suffering all the repressions and prohibitions which that requires. That steps can be taken, as a community, to limit the repressions and to deny those prohibitions which promise us survival at the price of our minor dignity means that all hope need not be abandoned. It is even possible that technology can provide some slight assistance.

The individual will always have to bargain with the culture. Technology, far from making a better bargain possible, has made it all the more difficult. Freud's God—reason—would be served in a not inconsiderable way if it could only be seen that there can be no perfect outcome to the bargaining. There can only be, at best, a life of modest satisfactions. Among these, glints of altruisms, the transcendence which can come through a joining of human life for the sake of working against the common enemy of nature, the body and other human beings, and a controlled attempt to use technology to take us one tentative step forward, count for something.

## NOTES

1. A splendidly balanced defense of technology and its future possibilities can be found in René J. Dubos, "Humanizing the Earth," *Science*, Vol. 179 (February 23, 1973), pp. 769–72.
2. Stuart Hampshire, "Morality and Pessimism," *The New York Review of Books*, Vol. XIX (January 25, 1973), pp. 31–32.
3. Philip Rieff, "Fellow Teachers," *Salmagundi*, No. 20 (Summer–Fall 1972), pp. 35–36.
4. *Ibid.*, p. 69.
5. Colin M. Turnbull's study of the Ik, *The Mountain People* (New York: Simon and Schuster, 1972), is an exceedingly disturbing voyage into the ethics of survival. "The Ik have 'progressed,' one might say, since the change that has come to them has come with the advent of civilization to Africa and is therefore a part of that phenomenon we so blandly and unthinkingly refer to as progress. They have made of a world that was alive a world that is dead, a cold, dispassionate world that is without ugliness because it is without beauty, without hate because it is without love, and is without any realization of truth even, because it simply is. . . . The Ik have successfully

abandoned useless appendages, by which I refer to those 'basic' qualities such as family, cooperative sociality, belief, love, hope and so forth, for the very good reason that in their context these militated against survival. . . . they have replaced human society with a mere survival system that does not take human emotion into account. . . . Such interaction as there is within this system is one of mutual exploitation. That is the relationship between all, old and young, parent and child, brother and sister, husband and wife, friend and friend" (pp. 289–90).

# Index

# INDEX

INDEX

## INDEX

## INDEX